COGNITION, SEMANTICS AND PHILOSOPHY

PHILOSOPHICAL STUDIES SERIES

Founded by Wilfrid S. Sellars and Keith Lehrer

VOLUME 52

COGNITION, SEMANTICS AND PHILOSOPHY

Proceedings of the First International Colloquium on Cognitive Science

Edited by

JESÚS EZQUERRO

and

JESÚS M. LARRAZABAL

University of the Basque Country,
San Sebastián, Spain

Kluwer Academic Publishers

Dordrecht / Boston / London

ISBN 0-7923-1538-3

Published by Kluwer Academic Publishers,
P.O. Box 17, 3300 AA Dordrecht, The Netherlands.

Kluwer Academic Publishers incorporates
the publishing programmes of
D. Reidel, Martinus Nijhoff, Dr W. Junk and MTP Press.

Sold and distributed in the U.S.A. and Canada
by Kluwer Academic Publishers,
101 Philip Drive, Norwell, MA 02061, U.S.A.

In all other countries, sold and distributed
by Kluwer Academic Publishers Group,
P.O. Box 322, 3300 AH Dordrecht, The Netherlands.

Printed on acid-free paper

Printed in the Netherlands

CONTENTS

ACKNOWLEDGMENTS

We would like to thank Kluwer Academic Publishers for their patience in helping to put this volume together. We also would like to thank our colleagues in the Department of Logic and Philosophy of Science of the University of the Basque Country for helping to organize the Colloquium. Financial support came form the Ministerio de Educación y Ciencia (Madrid), Departamento de Educación (Gobierno Vasco), Diputación foral (Guipuzkoa), Kutxa Guipuzkoa-Donostia and the University of the Basque Country. Many thanks. Finally, thanks also to our friend Paul Horwich. He helped to contain our linguistic bugs.

INTRODUCTION

THE PLACE OF PHILOSOPHY IN COGNITIVE SCIENCE

During the last few years, many books have been published and many meetings have been held on Cognitive Science. A cursory review of their contents shows such a diversity of topics and approaches that one might well infer that there are no genuine criteria for classifying a paper or a lecture as a contribution to Cognitive Science. It is as though the only criterion is to have appeared in a book or in the programme of a meeting in whose name or title we can find the expression "...Cognitive Science" or something like that. Perhaps this situation is due to the (relative) youth of the field, which is seeking its own identity, still involved in a process of formation and consolidation within the scientific community; but there are actually deep disagreements about how a science of the mind should be worked out, including how to understand its own subject, that is, "the mind." While for some the term makes reference to a set of phenomena impossible to grasp by any scientific approach, for others "the mind" would be a sort of myth, and the mental terms await elimination by other more handy and empirically tractable terms. Still, some progress has been made and nowadays there seems to be some tacit agreement concerning what a suitable cognitive science ought to do, that is, to study "the mind" without confining itself to being a science of the brain, but also to do this job in a tangible, manageable and as exact as possible way.

No doubt, philosophy has had a lot to do with this state of the art, both in the good and in the bad features. Frequently it is claimed that philosophy does not bring forth results, at least in the way that one of the empirical sciences does. However, what philosophy actually has done, and should do, is to frame topics, problems and questions, to offer views, to analyze approaches, to point out what can be important, telling why.[1] And in doing all these things, philosophy has sometimes made mistakes, but has had successes as well. The mistakes have come out mainly, in our view, when philosophy has neglected its role of metadiscourse, trying to supplant and, in some cases, to exclude, the substantive research relevant to accounting for the cognitive phenomena worked out by scientific disciplines such as neurobiology, artificial intelligence, psychology, and so forth. Fortunately, nowadays we are far off from those philosophies of mind and philosophies of ordinary language which tried to govern the cognitive kingdom as a dictatorship, while linguists, psychologists, artificial intelligence researchers and neuroscientists were doing their jobs. Today, as it should be, philosophy follows, having a lot of things to say about what the particular sciences accounting for the several features of cognition are doing. The development of these sciences, together with conceptual reflection on the links between them and the proposal of conceptual frames, is conforming the discipline. Notwithstanding, philosophy has some things to do as well, at least from the point of view of a field of knowledge traditionally worked out within philosophical forums, which is of a central importance for studies of cognition, namely, semantics, both formal semantics and non-formal or philosophical semantics.

If there is an essential problem in cognitive science it is the problem of knowledge representation. Around this issue turns almost all that can be done or claimed within the discipline, insofar as its subject is information processing systems, that is, systems which handle information, acquiring and transforming it. In a nutshell, systems whose behaviour relies upon the information they manage. What shape should the represented information have? Should it be of a linguistic sort, or not? Which cognitive or computational architecture is the most adequate one? Which languages or linguistic systems are the most suitable ones for speaking about representations and about what representations allow us to do (for instance, reasoning, understanding, language performance, and so on). All these are hot questions nowadays, to whose answers the following papers are trying to contribute.

The present volume contains some of the contributions presented in the International Colloquium on Cognitive Science (ICCS-89) which was held in San Sebastian in May, 1989. The meeting was organised by the Department of Logic and Philosophy of Science of the University of the Basque Country. Both the design of the congress and the title of the volume—*Cognition, Semantics and Philosophy*—responds in part, to the interests of the organising institution, or rather, to the view that we, researchers with an academic training of a logical-philosophical sort, have about the cognitive science field in the sense above mentioned. This is the reason why semantics, understood in a wide sense, occupies a central place in the title, as a bridge between cognition and philosophy. This does not entail that the presented contributions are only from philosophers—more or less half of them are contributions from psychologists, linguists or artificial intelligence researchers —but that we are trying to point to a common area wherein the interests of the several trends working on cognitive science converge.

A pretty similar idea is the one proposed by F. Le Ny in his paper "*Cognitive Science and Semantic Representations.*" The author proposes the concept of "semantic representation" as the unifier notion for the different disciplines which come together in cognitive science. On the one hand, "knowledge representation" constitutes a well strengthened field in artificial intelligence, whereas the mental or natural representations which occur in individual brain-minds are investigated by cognitive psychology though there exists obvious differences concerning the goals and methods in both fields. On the other hand, if we accept that the description of the representational states of the agents must be made at the "knowledge level," such a level will be common to psychology and artificial intelligence. In the end, the heuristic resources for modelling in artificial intelligence come to a great extent from the intuitive observation of how we, human beings, do represent knowledge to ourselves. And reciprocally, when, from psychology,models are proposed for representing knowledge, it becomes unavoidable to use conceptual tools from logic and artificial intelligence. Furthermore, perhaps the main procedure for testing such models is just to implement them in a computer. The analogies and common areas can become extensive in a more fine grained way if we look at the distinction between the two main kinds of representations, that is, long term *versus* transient representations. This distinction finds its correlates in artificial intelligence with the lexicon of the data bases and the activation or instantation procedures of the elements of such bases. In a nutshell, the author's claim is that cognitive science with all disciplines who participate in it, should be connected around the psychological notion of "semantic representation."

V. Sánchez de Zavala, in the first part of his paper entitled "*On the Study of Linguistic Performance*," proposes a very different view of the way research in cognitive science should be articulated. Instead of proposing a central notion from which to establish a hierarchisation and location of the different approaches or work lines, the author points to the need to determine some central goals or tenets which would guide research within the several disciplines. In this author's view, one of these great topics should concern the theoretical status of linguistic activity. Several reasons favor this choice: it is a central issue in cognitive research given that it is a cognitive subject in the strongest sense, that is, the study of the mental activity which occurs while, and perhaps through, the use of language. After carrying out a detailed analysis of almost the entire field of linguistic performance, both in its productive and its receptive features and centering himself mainly on the case of oral-auditive speech, the author proposes, in the second part of his paper, that the most suitable way of developing a general cognitive theory accounting for the cognitive processes accompanying linguistic performance, would be a pragmatic approach, specifically, an approach along the lines sketched by Sperber and Wilson's Relevance Theory. Nevertheless, the author emphasizes serious limitations concerning what has been worked out within this approach. Among these limitations we can find its overly restricted scope of application and some difficult problems hidden behind its apparent explanative simplicity. However, the most serious problem which affects this approach is the apparent impossibility of applying it to the other great side of linguistic performance besides speech understanding, that is, the phenomenon of speech production. This last limitation makes necessary, the author believes, a full revision of Relevance Theory. To this end he proposes to attend to recent developments in action Theory and to research on modelling knowledge representation, for we can find in these fields very promising resources with which to mend the main deficiencies of Relevance Theory.

This interest in the study of language use, but linked mainly with the problem of language representation, is shared by T. Espinal. His paper "*On the representation of linguistic information*" deals with the problems which arise when we intend to describe formally the information carried by the so-called parenthetical or disjunct constituents; that is, the linguistic fragments which are not syntactically integrated within the structure of their host sentence. Their inability to solve these problems raises questions, in the view of the author, about the validity of several philosophical and linguistic approaches, such as the theory of Speech Acts or current grammatical theories which are compositional and assume the modularity hypothesis. Instead, with the aim of accounting adequately for the information of parenthetical constituents, the author proposes to use multidimensional syntactic structures, that is, sets of independent syntactic structures linked by means of different mechanisms whose synchronization hinges on several restrictions and principles of a syntactical, conceptual and pragmatic sort.

Between the representations and their use we can find the main topic of F. Verdejo's paper "*User Modelling in Knowledge-Based Systems.*" The author analyzes several problems that have arisen in the construction of computational models for interactive communication, for instance, systems composed of a computer system and its user. The point at stake then will be to obtain enough information about the user with the aim that effective communication can take place, and obviously, to find forms for representing such knowledge in such a way that it can be acquired and handled effectively. The consequences of this line of work on knowledge representation and its use are pretty obvious, and its development intersects with almost all areas which converge in cognitive science, that is, artificial intelligence, psychology, computational linguistics, epistemology, the sciences of education and so forth. The reason is that in accounting for interactive communication we encounter all the basic ingredients of intelligence: knowledge, learning, articulation and communication.

If knowledge, or its representation, is used, it must be acquired as well, and this feature is of basic importance when we try to analyze the relations between our representations and the world which, seemingly, they represent. No doubt, this is one of the hardest problems to be accounted for in cognitive science. The reason is that it raises new problems which do not arise in the mere representation. In accounting for the representations usually the research begins, with some exceptions, by taking for granted their linguistic shape. And in that case, the acquisition of linguistic representations from inputs and resources which do not have such a linguistic shape is pretty problematic. Furthermore, it seems plausible to think that at least part of the content of our representations do not have a linguistic character. According to C. Peacocke's paper

"*Anchoring Conceptual content: Scenarios and Perception,*" this is the case for the representational content of perceptual experiences. This author, in a work of special relevance for the problem of anchoring conceptual content in some non-conceptual content level, maintains the thesis that a representational content linked to perceptual experience has to be identified by specifying which ways of filling out the space around the perceiver are consistent with the soundness of that content. Thus, content presupposes a spatial *type*, and then this element affects the soundness of representational content by instantation, by the real world, of the spatial type which is given by the representational content. After characterising carefully his proposal, the author applies it to the explanation of some cognitive phenomena, especially those related to human action, especially spatial reasoning.

Along the same lines, but from a different perspective, L. Aguado in his "*Animal Cognition and Human Cognition: A Necessary Dialogue*" insists on the need for mutual acknowledgement of the close contact between research on animal cognition and research on human cognition. On the one hand, the fact that human beings share a great deal of their cognitive structures with animals and, on the other hand, the great number of theoretical and empirical results obtained by cognitive psychology which are applied to research on animal cognition, make essential the dialogue between both fields in order to get a general theory of cognitive systems. The advantages of this exchange would consist in situating human cognition in a much wider biological perspective, and then the use of experimental techniques created in the field of animal cognition could contribute to a better understanding of the neurobiological basis of cognition.

From a more normative than descriptive point of view, (though one must take into account the hard difficulties of distinguishing between these two perspectives when we are talking of knowledge representation and its dynamics), D. Edgington, in her "*ChangingBeliefs Rationally: Some Puzzles,*" treats the problem of rationality in changing beliefs. More specifically, with the assumptions that belief change admits degrees, and that belief degrees form a probabilistic structure, she focuses on the problem of how beliefs must be revised when new information arises. Of course, following these assumptions, the author defends, and proposes, a probabilistic frame for treating the problem. And she does so, not only because of the above mentioned assumptions, but especially because she thinks that previous treatments of the problem—simple conditionalization and generalized conditionalization (or Jeffrey's conditionalization) —do not cover all the interesting cases, since they say nothing about the situation when the change of probabilistic judgments is motivated by information that is itself conditional. For the treatment of these cases she proposes a new principle: the independence of conditional probabilities and their "antecedents." Even if this principle sometimes conflicts with the principle of simple conditionalization, the author considers that the cost of abandoning it is too high, and it is preferable to maintain it and explore ways to solve the paradoxes this principle produces.

Works oriented to the characterization of memory had, in their origin, a clearly psychological motivation. However, it has been seen recently how this field, which is central in the study of representation, intertwines directly with semantics, pragmatics, artificial intelligence, epistemology and linguistics, among others. Two examples in this volume of that interdisciplinary stand-point are the contributions by T. Myers et al. "*Partiality and Coherence in Concept Combination*" and by K. Stenning "*Modelling Memory for Models.*" The first one is dedicated to the analysis of lexical representation problems and coherence. After noticing the existence of an evident tension between semantic flexibility phenomena and specificity, they also note that this tension raises more difficulties to be solved in some treatments than in some others. Specifically, from the stand point of sense selection, the authors think that the difficult relationship between semantic flexibility and specificity produces unsolvable problems, since the strong version of this perspective is affected by a combinatorial explosion of the interpretation associated with complex expressions, and its weak version, even if it diminishes this problem, trips over the problem of coherence. The authors think that these problems can in large part be solved if the treatment by sense generation is used. The adoption of this perspective offers reliable clues for investigating how lexical meaning is represented. The authors explore this route, studying with several examples the treatment of coherence and combination of concepts. Perhaps the main consequences of the sense generation approach mentioned by the authors is that lexical concepts have to be considered as partial ones, that the theories chosen by agents in their cognitive activities have a strongly local nature, linked to the immediate situation, and that is why they choose among the use of one or another theory in virtue of their informational needs, and that when they do it, they are adopting different perspectives.

K. Stenning's paper, also concerns memory, although from a quite different point of view. Specifically, it shows the results of a research program aimed at the problem of how models are represented during and after text comprehension. The paper begins with some remarks about the importance of implementation in cognitive science, claiming that the type of computational architecture in which we express our theories determines to a great extent the type of phenomena we choose to work out. In this line, and assuming that, given the resource limitations exhibited by human beings in their cognitive activities (such as reasoning and memory), Von Neumann machines do not offer us the most adequate architecture to copy with these strengths and weakness, the author welcomes connectionist architectures as an important new tool for a psychologist interested in reasoning and memory. Really, his aim in this paper is to show that this new perspective brings to the center of attention the binding problem in memory, affirming that "a solution to this problem can explain features of human reasoning which have been claimed to show that human reasoning has no relation to logic." Further, if these explanations are correct, the author believes that human reasoning "is just a straightforward implementation of a classical logic in a system with a very different working memory architecture than that of a Von Neumann computer." Starting from these views, the paper goes on to describe the results of the program above mentioned, perhaps the most important one being that the analysis of memory errors "reveals that the underlying representation of individuals is distributed through a set of features representing existential facts about the whole ensemble of individuals described." Interestingly, a PDP simulation of these representations explain human error patterns. The paper ends with a reconsideration of Johnson-Laird's theory of syllogistic reasoning.

In some sense, this later paper could be considered a new step in what is usually called today "the architecture ware," which confronts classical and connectionist approaches. From a more philosophical perspective, M. Davies's paper *"Aunty's Own Argument for the Language of Thought,"* is devoted directly to this hot topic. In a nutshell, what M. Davies's paper aims to point out is an incompatibility between our image as thinking persons, that is, the image provided by folk psychology, and the one derived from connectionists approaches. The paper begins with a discussion of the philosophical background of Aunty's position (remember, Fodor's uncle). The author, thinks of her as a rather Oxonian Aunty, say with somewhat Wittgenstenian tendencies, but fundamentally as a neo-Fregean. As such, "she holds it is a deep truth that thoughts are structured: thoughts have concepts as constituents, and these constituents can be recombined into other thoughts." So, the greater part of the paper is devoted to carefully analyzing and making clear what Aunty's argument amounts to. This analysis reveals the tension above mentioned between our commonsense conception of our mental life and the connectionist program for modelling cognitive processes. The author's argument parallels pretty closely to the one recently proposed by Ramsey, Stich and Garon (1990),[2] and could be summarized, as in this case, as proposing the following conditional claim: If connectionist hypotheses turn out to be right, then folk psychology must be false. That is, the author argues, as Aunty does, that "networks do not exhibit syntax and causal systematicity of process: the commonsense scheme is committed to syntax and causal systematicity of process; therefore connectionism is opposed to the commonsense scheme." However, the author is cautious about the prospects of this *a priori* argument. After all, we cannot prejudge the outcome of research in cognitive science. On the other hand, in so far as there are some insights about ow to reduce the tension between the two views,it could be that the requirements of syntax and systematicity would accommodated in some higher level of description of our cognitive architecture.

The interaction of semantic and pragmatic aspects in the interpretation of attitude reports constitutes the central issue in D. Quesada's paper *"The Labyrinth of Attitude Reports."* Taking off from some assumptions about what mental states are, within a naturalistic but non-reductive view, that is, affirming that they are physical states semantically described, and noting that this sort of description is indispensable if we want to make sense of "how they cause the actions they do cause," the author describes briefly the semantic framework concerning names, definite descriptions, personal pronouns and demonstratives, in which he will place his discussion. The main bulk of the paper analyzes a recent proposal due to Crimins and Perry about the relation between the semantic content of mental states and the content of the embedded sentence in a report about such mental states. These authors affirm that a belief report consists in an existential claim of the sort "there is a belief of an agent, with such semantic content and meeting certain condition," it being part of the semantic content of the report that the belief meets the condition. So, a belief report is a claim about the subjective notions of the agent. Problems arise, from the author's view,if we take into account that this proposal makes the attitude reports devoted to the perspective of the agent, and it will get us in trouble when the reports are not so focused. In fact, the author claims, there are lots of situations in which is very unreasonable for a speaker (or hearer) to believe that she should be taken as intending to describe the other agent's state of mind. For this reason, he proposes to introduce a distinction between two kinds of attitude reports, a distinction which parallels the well-known *de dicto/de re* distinction, although with a slight different meaning.

To conclude, we hope to have given in the present volume a representative view of work in cognitive science, which will be interest form a philosophical point of view. After all, knowledge, and of course, knowledge representation, learning and behaviour, have been among the oldest topics in philosophy ever since its Greek origins. Nowadays, the entering onto the scene of cognitive science is changing, we believe, in a paradigmatic sense, the course of inquiry in this field. No doubt, some philosophers will claim that philosophy, and specially epistemology or theory of knowledge, are essentially normative projects. It could be so. Moreover, even if one does not favor the Quinean tenets of naturalizing epistemology, and even if one is skeptical concerning the possibility of reducing the normative notions of meaning, truth, reasoning and so forth, to descriptive ones, or of eliminating them, the work done in cognitive science will continue being of undeniable interest for philosophy. Moreover, there remains semantics. No account of knowledge representation, language, learning and use can avoid dealing with semantic issues. After all, semantics itself is essentially concerned with representations, and because of this character, semantics is called on to play one of the main roles in establishing the links between cognition and philosophy.

NOTES

1. For a point of view pretty similar to this one concerning the role of philosophy in Cognitive Science, see D. Dennett's (1987) "Setting Off on the Right Foot" in his *The Intentional Stance*, Bradford, M.I.T.

2. "Connectionism, Eliminativism and the Future of Folk Psychology" in J.E. Tomberlin (ed.) *Philosophical perspectives 4: Action Theory and Philosophy of Mind*, Atascadero, CA.: Ridgeview Publishing Company, pp. 499–533.

CHAPTER 1

ANIMAL COGNITION AND HUMAN COGNITION: A NECESSARY DIALOGUE

Luis Aguado Aguilar
Universidad Complutense de Madrid, Spain

I. INTRODUCTION

Psychology has held through its history an ambiguous attitude towards animal behavior. Researchers who have chosen some animal different than man as their experimental subject have often sought to justify themselves, emphasizing the potential benefits that the understanding of animal behavior could eventually have for understanding human behavior or even for alleviating people suffering from psychological distress. During the behavioristic era, animals, especially some privileged species, were the usual subjects of many psychological experiments and the classical theories of learning were based almost exclusively on evidence gathered from animal experiments. On the other hand, the technology of behavior modification was, at least in its beginnings, an attempt to apply the general principles of classical and instrumental conditioning to the treatment of human clinical problems. However, there were not many explicit attempts at providing animal models of human psychopathological syndromes; rather, parallelisms between laboratory phenomena and clinical observations would arise accidentally and would be later exploited (a case in point is the learned helplessness phenomenon in animals, that has been considered as a model of human depressive behavior (Seligman, 1975).

Evolutionary arguments were not common among traditional behaviorists and when they were purportedly looking for general rules of learning (that is, conditioning) they did not explicitly state why should those laws exist. Actually, the claim that some general processes of simple associative learning might have been shaped in the course of the evolution of separated animal lineages under the pressure of some features common to all environments (convergent evolution) has become only recently a piece of the defensive rethorics of animal researchers. For example, it has been said that as the events occurring in any environment are causally related, evolution has led to the selection of similar mechanisms of detection of those relationships in very different species (Dickinson, 1980).

The comparative psychology of intelligence (see, for example, Bitterman, 1975) resorted to the tactics of comparing the performance of several species in standard laboratory tasks that supposedly forced the animal to display its cognitive resources. The goal of this line of research was to set up a phylogenetic scale of intelligence and its failure was due in part to an excessive reliance in a simplistic view of evolution as linear progress and in a conception of intelligence as a global capacity that different species might possess in a different degree (for classical critiques of this approach see Hodos and Campbell, 1969 and Lockard, 1971).

II. CHARACTERIZATION OF COMPARATIVE COGNITION

Cognitive psychologists do not use to show much interest in animal behavior and are often sceptic when one tries to persuade them that doing animal research is not necessarily a waste of time. The scant popularity of animal research among cognitive psychologists stems perhaps from the identification of animal research with behaviorism, conditioning and stimulus-response theories (if this identification were correct, there would be good reasons for rejection). However, out of the traditional animal learning laboratory a very active area of research has recently developed under the heading of comparative or animal cognition, whose theoretical foundations and research tactics are not substantially different from those underlying standard cognitive psychology.

The object of comparative or animal cognition is the experimental study of the cognitive processes underlying the adaptation of animal behavior to the environment. Perception, attention, learning and memory, communication, problem solving, and even consciousness and intentionality are interest topics for comparative researchers. To this end, they have adopted a wide range of methods derived from the technology of animal conditioning and learning, that allow "questioning" animals and making inferences about the course and content of their mental activities.

The conceptual framework of comparative cognition is that of human cognitive psychology, that is, what may be loosely termed the information processing approach. Comparative researchers consider, as human cognitive researchers do, that the main function of the brain is the processing of information given by sensory organs and thus that the goal of cognitive theories is to specify the processing operations carried out by the brain and the content of the symbols it works with, that is, mental representations. The study of animal cognition rests on the conception of cognition as an activity of a biological system that, as such, has an adaptive function and is subject to the action of natural selection. From the standpoint of comparative cognition, a research program that does not overlook the biological nature of cognition should have as its ultimate goal the building of a general theory of the structure and workings of cognitive systems and its use in the control of behavior.

III. COGNITIVE MODULES AND EVOLUTION

The aim of this paper is to discuss some theoretical and empirical developments of current cognitive psychology that might serve as guidelines for animal cognitive research. Given that the human cognitive system is the best known for us, due both to our direct subjective experience and to scientific evidence, it is a logical starting point for wondering about the properties of the cognitive systems of other species. At the same time, the arguments to be developed here are aimed at placing human cognition in an evolutionary, non-anthropocentric, perspective. Some of my arguments are openly speculative and do not mean anything but prompting reflection on issues about which we are far even from asking the right questions.

One of the recurrent themes of this paper will be that the scant interest of cognitive psychologists for animal cognition is based on a too restrictive conception of cognition. Since, as I will try to show, there is abundant evidence challenging that conception, one of the obstacles hindering the interchange of data and ideas between researchers of animal and human cognition should finally be torn down. The progress of the sciences that study behavior and cognition, whether it is from a behavioral, computational or a neurophysiological perspective, depends on an effort of mutual collaboration and animal research might have an important role to play in this respect.

Reserves concerning the relevance that the study of animal cognition might have for understanding human cognition seem to be based on the identification of cognition with language and consciousness. This identification may take two forms: that cognitive activities are only those conscious mental activities of a subject who uses language; or else, that even being only a part of what may be considered cognitive, language and consciousness permeate to such extent all our cognitive activities that human cognition is radically different from the cognitive processes of other species. There are arguments, both empirical and theoretical, that challenge these views.

I shall begin with a theoretical argument based on the modular theory of mind put forward by Fodor (1983). Though Fodor's arguments refer specifically to human cognition, they might be interpreted in a more general sense as a theory about the structure and basic properties of cognitive biological systems. The core of the modular theory is the distinction between "vertical" and "horizontal" mental faculties, briefly, between perception and "higher" cognitive processes. One of the features differentiating between these two kinds of faculties or cognitive capacities is that vertical capacities (i.e.,perceptual systems and, for Fodor, language too) are based on independent, encapsulated, modules or subsystems, in the sense that they are not affected by "higher" processes (for example, expectancies, intentions, beliefs or rational deductions of the subject). However, as the vast array of empirical evidence on the workings of perceptual systems show, in spite of its independence from higher processes perception involves a complex processing of sensory input that is carried out in an automatic and non-conscious way. According to this view of mind, it can be said that even in man a significant part of cognition is carried out with independence of those processes traditionally considered as higher cognitive activities and that the identification of cognition with consciousness leads to a partial and distorted view of cognition.

The modular view of mind has at least two implications for the comparative study of cognition:

1) The idea of cognitive systems working without a significant supply of what are usually called "higher" mental capacities becomes more plausible. That a processing activity is automatic and non-reflexive does not mean that this activity is cognitively uninteresting or irrelevant and, because probably a big part of the mental activity of animals is of this kind, animal cognition becomes an interesting topic.

2) If the activity of modular systems is not substantially modified by higher processes, there might be important similarities between the mechanisms governing these systems in different species, no matter their relative endowment of higher mental abilities. The modular view suggests where there will be more probable to find similarity or divergence among species. It should be more probable to find similarities (general properties) in the structure and functioning of modular systems and qualitative divergencies in the processes served by higher faculties. Curiously enough, traditional comparative psychology focused precisely on comparing different species in problem solving tasks aimed at tapping the "highest" mental capacities of the animal. In doing this, comparative psychology seems to have implicitly recognised that perception was not a good place to start if what one wished to find were qualitatively differences among species (after all, perception is not intelligence).

The modular view has also made its way through the neurosciences, specially neuropsychology. In this context, the term "module" is applied in a broad sense and is not necessarily defined by the same features as fodorian modules, referring rather to dissociable neural subsystems with specific functions. Though I shall return later to this point, a well known example is the conservation of general capacities of reasoning, intelligence, memory and attention in aphasic patients (see, for example, Kertesz, 1988). An area in which the modular or componential view is specially prominent is the neuropsychology of memory; the view of the brain as a large assembly of independent memory systems or modules usually working in interaction and each devoted to the processing or storing of different features of the same target event is based on the results of numerous studies showing selective impairments produced by localized brain lesions or disfunctions (see, for example, Squire, 1987).

The modular conception combined with some evolutionary concepts, specially those of conservation in evolution and convergent evolution, helps to give some theoretical foundation to the research program of comparative cognition. Biological evolution acts usually in a conservative way, not giving rise to new structures if a new function can be performed by an already existing structure. If this is true of the morphology and functioning of organic structures that allow, for example, locomotion, flight or digestion, it is not easy to see why brain evolution should have followed a different strategy. Being as it is an aggregate of components and not an undifferentiated structure, it is probable that the brain has evolved along each phylogenetic line through additions and reorganizations of already existing systems rather than through radical qualitative changes. The idea of conservation in evolution, together with the modular view of the mind and brain and the empirical data of neuropsychology, make hardly defensible the conception of the human mind as an strange mutant radically different from the minds of all the other species. Moreover, there are probably some common demands placed by different environments upon cognitive systems, concerning both perception and the storing of information and its use in the control of behavior, which might have led by convergent evolution to the development of similar processing mechanisms even in phylogenetically distant species. As will be discussed in the next section, the discovery of such similarities is precisely one of the main goals of the comparative approach to cognition.

IV. TWO GOALS OF COMPARATIVE RESEARCH: GENERAL PROCESSES AND EVOLUTIONARY SEQUENCES

The arguments put forward in the preceding section may serve as guidelines for comparative research. One of the goals of modern research on animal cognition is the identification of cognitive systems serving similar functions in different species and the comparative analysis of their mechanisms. This tactic is aimed at finding out if a behavioral ability present in different species is based on a cognitive system governed by common mechanisms or if there are variations in mechanisms among species. An example of this is the way in which different species deal with the problem of orientation in space. Though several orientation mechanisms, based on the processing of very different kinds of cues, have been discovered (detection of terrestrial magnetism, orientation guided by sun position in the sky, use of olfactory cues...(see, for example, Keeton, 1974)) a widely spread mechanism of orientation seems to depend on the visual identification of specific landmarks and its representational base seems to be similar for different species.

In laboratory studies carried out with different species (see, the experiments of Cheng and Gallistel (1984); or Suzuki, Augerinos and Black (1980) with rats; Fabrigoule and Maurel (1982) with foxes; Kamil and Balda (1985) with birds; Gould and Gould (1982) with bees; or Menzel (1978) with chimpanzees) the representational basis of orientation in space seems to be a cognitive map containing specific information about different landmarks and its relative positions in space. This information may be used flexibly enough as to allow the animal, for example, to take a new route to arrive to an already visited place. The basic strategy of the experiments on cognitive maps involves introducing variations in the position of the cues after original training in an spatial problem. By looking at which variations (e.g., distance between cues, rotation of the cues keeping constant their relative positions, and so on) disturb performance and which ones leave it unaffected, it is possible to know, first, if the representation of the environment in the animal's memory is of the map kind and, second, which of the specific features defining the relative position of the cues (distance, direction) are represented in the map.

New examples of the search for general properties of cognitive systems may be found in the comparative research on perception. For example, with the aid of discriminative training methods it has been observed that perception of visual forms by monkeys and pigeons show contextual effects, as human visual perception. In one example of this effects, recognition of a simple feature was made easier by its inclusion in a complex figure even if the rest of the features in the figure was redundant to the solution of the discrimination (Blough, 1984; Schrier, 1985). This result suggests that animal perception may follow organizative principles similar to those usually found in man.

One final example of the generalist approach to comparative cognition is the research program on animal memory carried out by Anthony Wright and his co-workers. They have used mainly a serial learning paradigm with lists of several items that the animal has to memorize. It is well known that humans show serial position effects of primacy and recency (best memory for the initial and last items) when asked to remember the items of a previously presented list of words or paired associates. Serial position effects have been usually discussed in the context of the traditional distinction between short and long term memory systems and have been considered as supporting this distinction (e.g., Glanzer & Cunitz, 1966). Serial position effects identical to those found in man have also been shown in pigeons and chimpanzees in tasks in which the animal has to recognize which of several complex visual stimuli serially presented were in a list of images to which he has previously been exposed (Wright, Santiago, Sands and Urcuioli, 1984). Specifically, correct identification is more probable for the initial and last elements and errors of identification are higher for the intermediate ones. Moreover, increasing the delay between the list and the memory test eliminates the recency effect (Wright, Santiago, Sands, Kendrick and Cook, 1985). Though a behavioral similarity among different species under equivalent testing conditions does not necessarily imply an underlying common mechanism, it might be that these results reflect a true similarity in the working of the memory systems of such phylogenetically distant species as monkeys and pigeons.

The generalist strategy is not the only approach to comparative cognition. Another goal of comparative research is to describe possible evolutionary sequences giving rise to complex cognitive capacities. The basic strategy in this case involves the search for more elementary, prerequisite capacities, that might be present in species that do not show the complex one. Such a program of research can hardly come to definitive conclusions about the precise evolutionary origin of a complex cognitive capacity, but may at least give more or less plausibility to different accounts of its nature. This strategy might, for example, provide some clues on the nature of human language, specifically on the issue of whether it is a totally new and self-sufficient capacity or whether it is based on general cognitive capacities that are themselves independent from language.

Language allows us to assign names to classes of objects, events or situations. Although human beings use words to express concepts, it might be that concepts and categorization in general are not a consequence of the possession of a complex code to communicate with the others and express thinking, but rather that abstraction is a prerequisite for the use of language. The study of human development gives fundamental clues for understanding the origin of categorization and abstraction, but comparative animal research may also make significant contributions. Animal researchers have insistently tried to find evidence of non linguistic abstraction. The basic experimental paradigm is a variety of discriminative training in which the animal must discriminate between two large sets of stimuli, one comprising several exemplars of a class, as defined by the researcher (usually natural categories as "fish," "tree" or "people") and the other composed by stimuli not belonging to that class. During the course of training, the subject is reinforced for choosing or responding to the stimuli representing exemplars of the category.

Animals learn with striking facility to solve this kind of "concept discriminations," but this learning is not in itself a proof that the animal has formed a concept (he might simply have learned a series of associations between each stimulus and reinforcement or non-reinforcement). A more stringent test of concept learning is a transfer test in which the animal is confronted with two sets of new stimuli, one with new exemplars from the target category and the other with new non-category stimuli. It is expected that, insofar as the animal has solved the original problem on a conceptual basis (for example, abstracting the common features shared by all the members of the category), he will be able to transfer immediately this knowledge to the new stimuli, identifying them as members or non-members of the category. The animals tested on this paradigm usually show this transfer (e.g., Herrnstein, Loveland and Cable, 1976). However, the results of the experiments on concept formation and identification are plagued with difficulties of interpretation that are, by the way, representative of the many problems that the comparative study of cognition has to face. For example, it is very difficult to exclude an interpretation of the transfer to new exemplars in terms of simple associative processes and non-conceptual generalization between the training and test stimuli. Moreover, there is always the empirical problem of finding out what is the level of abstraction that the animal is using when identifying an specific stimulus as member of a class. In a study by Roberts and Mazmanian (1988) it was observed that while human subjects transferred easily from the original discrimination to new exemplars when the class was of a high abstraction level (animal versus non-animal) and had difficulties with less abstract classes (a specific kind of bird versus birds in general), monkeys and pigeons showed precisely the opposite pattern, showing more transference the less abstract was the target category. This result suggests that animals found easier to group together stimuli with high visual similarity while humans tended to overlook visual similarity, centering instead on more abstract similarities among stimuli.

V. CONSCIOUSNESS AND COGNITION

Conscious thinking is one of the features usually considered as differentiating human from animal cognition. The terms "consciousness" or "conscious thinking" will be used here as referred to different functions, as self-awareness (the perception of oneself as a differentiated "object," specially as different from conspecifics), the initiation of voluntary actions oriented towards some predetermined goal, metacognition (or the awareness of one's cognitive processes) or the deliberate search and retrieval of information from memory. If we take for granted that consciousness is a higher mental capacity, that is, a capacity that is only present in species with a complex cognitive system, the relation between consciousness and cognition can be conceived in two ways.

We may suppose that the evolution of higher cognitive capacities, as consciousness, involves necessarily a modification of the specie's whole cognitive system. From this point of view, the higher capacity would control and permeate simpler capacities in such a way that these would become governed by new mechanisms, different from the ones that govern the same capacity in species not endowed with the higher one. Pavlovian conditioning, for example, is a lower or simple form of learning, in the sense that it can be found in many species, even in invertebrates with a relatively simple nervous system, whose cognitive activities are probably very modest, at least for human standards. Man is, in spite of some hazardous claims (Brewer, 1972), one of the species in which this form of simple associative learning can be observed. However, some researchers of human classical conditioning claim to have repeatedly found that their subjects show acquisition of conditioned autonomic responses only if they are aware of the associative relation between the conditioned and unconditioned stimuli (e.g., Dawson and Schell, 1982).

Without going into details about how representative this finding may be, it can be taken as an example of the absorption or control of a lower capacity (simple associative learning) by a higher one (awareness).

There is an alternative and more conservative (in the evolutionary sense) way of conceiving the relations between consciousness and the remaining cognitive activities. Consciousness may be thought as an added capacity that would allow the subject a new way of accessing or, to some extent, controlling other cognitive activities, but without substantially modifying their operation. Although the conception of consciousness as absorbing other cognitive activities does not necessarily invalidate the research program of comparative cognition, it gives some strength to the arguments of those who are sceptical about its relevance to the understanding of human cognition. The second alternative, that is, the idea of consciousness as an added capacity, facilitates in any case the task of comparative researchers and allows us to cherish more confidence in their success. But which of the two alternatives has a greater empirical support? There is, to my opinion, substantial evidence coming from different areas of research that seem more compatible with the hypothesis of consciousness as an added capacity. This evidence shows that many cognitive activities can be dissociated from their conscious experience. Two tentative conclusions can be derived from this evidence: first, that a significant part of cognitive processing proceeds independently from consciousness and, second, that even some activities that usually take place under the influence of consciousness may follow an effective, though hidden, course, in the absence of any awareness. I will briefly discuss some of this evidence.

During the last years there has been a heated debate in cognitive psychology about the role of unconscious or automatic processing. The impetus for the rising tide of the interest in this topic came specially from a series of experiments by Anthony Marcel on the effect of visual masking on word recognition (Marcel, 1983). Through the manipulation of SOA (stimulus onset asynchrony), Marcel showed, for example, that as the interval between the target, visually presented word, and a following mask stimulus is reduced, there is a decrease of, first, the accuracy with which the subject detects the word, second, of the accuracy with which he judges the graphic similarity between test and target words and, finally, a decrease of the accuracy with which the semantic similarity of the test words with the target word is judged. What is more striking is that the order in which accuracy decreases is precisely the opposite to which one would expect if processing of the word proceeded from its initial conscious detection to its graphic identification and, finally, to its semantic identification. The most important implication of this finding is that even a "higher" cognitive activity as semantic identification can be carried out automatically and independently from consciousness.

A different kind of evidence relevant for the problem of the relation between consciousness and cognition is provided by some recent neurophysiological studies that suggest that the initiation of motor voluntary acts might have in fact an unconscious origin. Libet (see Libet, 1985, for a review) has carried out a series of experiments on the relation between electrophysiological readiness potentials (ERPs) and the initiation of simple, seemingly voluntary, motor acts. ERPs precede normally the initiation of a motor act and, what is more surprising, in Libet's studies preceded in 400 msecs the moment in which the subject reported the intention to perform it. An interesting result of Libet's experiments was that his subjects were able to control the act (specifically, vetoing it) only during the 150 msecs following the report of the conscious decision to behave. In Libet's opinion, conscious decision, rather than really initiating voluntary acts would have the role of allowing or vetoing (a kind of gating control) performance of acts whose origin is unconscious.

Libet's account of the role of conscious decision suggests some speculations about different possible levels of voluntary control of motor acts in other species. At the most primitive level, motor acts would be initiated and performed in a totally "blind" way, released by antecedent stimuli without awareness of its execution. A higher level would involve awareness without the possibility of control or veto. This controlling function would be privative of the higher level, in which there would be a proper conscious control of action. This account may help us to see under a new light the distinction between voluntary and goal-oriented behavior. The orientation of a behavior towards an specific goal may be thought as the most primitive manifestation of purposive behavior; for example, when an animal has learned to expect some rewarding event, as food, following an instrumental action, there is no need to attribute the animal a conscious decision to perform the action (see Dickinson, 1988).

A system in which a motivational state (hunger, for example) increases the probability of any instrumental action which in the past has been followed by reward in a specific context may suffice to explain the behavior. The instrumental behavior would be oriented towards a goal, insofar as it is controlled by the expectancy of food in that context, but at the same time it would be inevitable, that is, not voluntarily controllable by the subject. Though the animal may be said to be "aware" of the consequences of his behavior in the sense that he has learned to expect a specific event after performing it, there are probably no grounds to say that he "decides" to behave in such way.

A topic that recently has aroused much interest among researchers of human memory is the distinction between explicit and implicit knowledge. This distinction is based on the dissociation sometimes found between the availability of some information or skill and the consciousness of such knowledge. The most widely discussed dissociation between explicit and implicit knowledge is found in amnesic patients, specially in those who have suffered a localized lesion in the temporal lobes. This kind of amnesiacs show a pronounced impairment on recall and recognition tests in which they are asked explicitly to retrieve some previously acquired information. However, when the availability of this information is tested through an implicit memory test not requiring conscious recollection, they often perform at a level similar to that of normal controls. Examples of tasks that allow to unveil implicit memory, showing conservation of information in amnesic patients, are word completion without instructions to remember, performance of motor skills and priming designs (Cohen and Squire, 1980; Jacoby and Witherspoon, 1982; Schachter, 1988). These results suggest that the deficit in amnesia is not one of storage of previously acquired information; rather, what seems to be impaired is the ability to access or retrieve this information in a voluntary or conscious way. This dissociation between memory and consciousness shows that knowledge may affect behavior even if the subject is not aware of it and is not able to accede to it deliberately.

Taken together, the evidence just discussed is consistent with the conception of consciousness as an ability separable from other cognitive capacities. For this conception, conscious thinking and awareness allow a certain degree of control and voluntary access to the output of different cognitive systems, specially those systems performing functions of processing of sensory information and storing of new knowledge and skills. A possible trend in the evolution of cognition might have been, precisely, an increasing flexibility in dealing with information stored in memory, leading for example to the development of explicit and voluntary retrieval strategies. But, apart from these voluntary strategies, there might be important similarities among the cognitive systems of different species. Following this reasoning, one line of research in comparative cognition would involve the search for common properties of implicit retrieval processes in different species as it is exemplified in several studies on the effects of context change on memory (Mackintosh, 1985).

After discussing the relations between consciousness and cognition one may ask whether any species different than man show some sign of awareness and intentionality. Though some students of animal behavior maintain that consciousness is a widely distributed capacity that can be found at many phylogenetical levels and that does not necessarily require a complex brain (see Griffin, 1981), the fact is that the only evidence suggesting some level of awareness and intentionality in animals comes from experiments with primates. For example, recognition of one's image in a mirror, often considered as an index of self-awareness (Gallup, 1975), has only been found in chimpanzees and orangutans (see the review by Povinelli, 1987). In other species, it has been found that the animal treats his mirror image as if it were another conspecific, even if he is able to use mirror cues to orient his behavior towards an item, such as food, that is not directly visible.

It is not surprising that it is in the context of social interaction, between conspecifics or with man, where it has been easier to find in non-human primates behaviors that might be interpreted as manifestations of consciousness, deliberation or even an incipient metacognitive ability. Life in a group with a relatively flexible social structure, for example one in which dominance relations vary from time to time, forces the animal to interact in complex ways with the other members, adapting his behavior to possible changes of status and anticipating what behavior will the others show. As some writers have pointed out (i.e., Dennet, 1983), one basis for predicting the behavior of another animal is attributing him intentions, expectancies and desires similar to those which one perceives in oneself. This kind of subjective inference requires, of course, that the animal is in the first place aware of at least some of his own mental states; moreover, he must have some knowledge of the relations between the other's overt behavior and the probable underlying mental states, or what have been called a "theory of mind" (Premack and Woodruff, 1978).

Several experimental paradigms have been proposed as a mean to study consciousness and intentionality in animals. Some are designed as situations in which the animal may get some benefit for himself giving some other animal an erroneous information. Though natural observation provides some evidence of "lying"during the social interaction among primates (Whitten and Byrne, 1988), there is also evidence of this kind of intentionality in more controlled settings. In a study by Woodruff and Premack (1979), a chimpanzee was confronted with two human experimenters, one playing the role of a cooperative partner and the other showing a deceptive attitude; while the first shared his food with the chimpanzee, the cheater kept to himself all the food that he found. When after this experience food was placed in one of two containers, out of the chimpanzee's reach, he reacted in a very different way depending on which of the experimenters happened to come into the room. While the chimpanzee behaved cooperatively in the presence of the cooperative experimenter, showing behaviors indicative of his knowledge of where food was hidden, he inhibited these behaviors in the presence of the cheater, sometimes even making gestures seemingly meant to confound the experimenter (for example, looking in the opposite direction to where the food was or pointing to the empty container). In another interesting study (Premack and Woodruff, 1978), chimpanzees watched a film in which a human actor struggled to solve a practical problem. Before the actor completed the action, the film was interrupted and the animal had to decide which of several alternatives represented the logical conclusion to the action depicted in the film. The fact that the chimpanzees chose the correct alternative suggests that they understood the behavior of the actor, realizing the goal he was pursuing. Results like those described in this section suggest that non-human primates may at least sometimes act intentionally on the basis of their predictions of the behavior of other animals, guided by a kind of subjective inference.

VI. CONCLUSIONS

I have stated through this paper some reflections about how data and theory derived from human cognitive research may be integrated with research on animal or comparative cognition. Though I have insisted in how human cognitive psychology may guide comparative research, I believe that both comparative researchers and cognitive psychologists may get some benefits from this integration insofar as both parts hold a receptive attitude towards the advances registered in each field. I will list, as a conclusion, the most obvious of these benefits.

1) The dialogue between human cognitive psychology and comparative cognition might help to orient cognitive researchers towards the study of topics relevant to the formulation of a general theory of the structure, functioning and use of cognitive systems. Human cognitive psychology offers well contrasted empirical and theoretical approaches that may be applied to the study of the cognitive systems of other species.

2) Comparative cognition may provide cognitive psychology with an evolutionary framework that allows to place human cognition in a broader biological perspective. In this sense, comparative research could help to test the generality of theories based on data from human research.

3) The development of new experimental techniques for the study of animal cognition may contribute to the advance of our knowledge of the neurobiological basis of cognition. A vast amount of neuroscientific research has been carried out with animals, but researchers have often resorted to very simple behavioral preparations that are not specially informative for those who are interested in cognitive processes. Comparative researchers have designed a vast array of experimental paradigms that make easier the analysis of such processes and that might help to develop suitable animal models for the study of the neurobiology of cognition.

4) Some techniques of comparative research aimed at the study of how the animals use their cognitive abilities in ecological contexts have given remarkable results concerning the psychological basis of natural behavior (see, for example, the studies on the memory capacities of food-storing birds reviewed by Sherry, 1988). Some cognitive researchers have expressed their concern about the ecological validity of laboratory research and have

urged the adoption of an ecological approach aimed at the study of cognition in natural contexts (Neisser, 1976). The study of the psychological basis of animal behavior in natural settings, or in laboratory situations reproducing some features of the natural environment, is surely a promising way of approaching the study of the relations between cognition and ecology.

REFERENCES

Bitterman, M.E.P. (1975) "The Comparative Analysis of Learning." *Science*, 188, 699–709.

Blough, D.S. (1984) "Form Recognition in Pigeons." In H.L. Roitblat, T.G. Bever and H.S. Terrace (Eds.) *Animal Cognition*. Hillsdale: Lawrence Erlbaum Associates.

Brewer, W.F. (1974) "There Is No Convincing Evidence For Operant And Classical Conditioning In Humans." In W.B. Weimer and D.J. Palermo (Eds.) *Cognition and the Symbolic Processes*. Hillsdale: Lawrence Erlbaum Associates.

Cheng, K. and C.R. Gallistel (1984) "Testing the Geometrical Power of An Animal's Spatial Representation." In H.L. Roitblat, T.G. Bever and H.S. Terrace (Eds.) *Animal Cognition*. Hillsdale: Lawrence Erlbaum Associates.

Cohen, N.J. and L.R. Squire (1980) "Preserved Learning and Retention of Pattern Analyzing Skill in Amnesia: Dissociation of 'Knowing How' and 'Knowing That'." *Science*, 210, 207–209.

Dawson, M.E. and A.M. Schell (1982) "Electrodermal Responses to Attended and Nonattended Significant Stimuli During Dichotic Listening." *Journal of Experimental Psychology: Human Perception and Performance*, 8, 315–324.

Dickinson, A. (1980) *Contemporary Animal Learning Theory*. Cambridge University Press.

........., (1988) "Intentionality in Animal Conditioning." In L. Weiskrantz (ed.) *Thought Without Language*. Oxford: Clarendon.

Fabrigoule, E.C. and A.D. Maurel (1982) "Radio-Tracking Study of Foxes' Movements Related to Their Home Range. A Cognitive Hypothesis." *Quarterly Journal of Experimental Psychology*, 34B, 195–208.

Fodor, J. (1983) *The Modularity of Mind*. Cambridge, Mass.: MIT Press.

Gallup, G.G. (1975) "Towards an Operational Definition of Self-Awareness." In R.H. Tuttle (Ed.) *Socio-ecology and Psychology of Primates*. The Hague, Netherlands: Mouton.

Glanzer, M. and A.R. Cunitz (1966) "Two Storage Mechanisms in Free Recall." *Journal of Verbal Learning and Verbal Behavior*, 5, 351–360.

Gould, J.L. and C.G. Gould (1982) "The Insect Mind: Physics or Metaphysics." In D. Griffin (Ed.) *Animal Mind-Human Mind*. Berlin: Springer.

Griffin, D. (1981) *The Question of Animal Awareness*. New York: Rockefeller University Press.

Herrnstein, R.J., D.H. Loveland and C. Cable (1976) "Natural Concepts in Pigeons." *Journal of Experimental Psychology: Animal Behavior Processes*, 2, 285–311.

Hodos, W. and C.B.G. Campbell (1969) "*Scala Naturae*: Why There Is No Theory Of Comparative Psychology." *Psychological Review*, 76, 337–350.

Jacoby, L.L. and D. Witherspoon (1982) "Remembering Without Awareness." *Canadian Journal of Psychology*, 32, 300–324.

Kamil, A.C. and R.P. Balda (1985) "Cache Recovery and Spatial Memory in Clark's Nutcrackers." *Journal of Experimental Psychology: Animal Behavior Processes*, 11, 1, 95–111.

Keeton, W.T. (1974) "The Orientational An Navigational Basis Of Homing In Birds." *Advances in the Study of Behavior*, 5, 47–132.

Kertesz, A. (1988) "Cognitive Function in Severe Aphasia." In L. Weiskrantz (Ed.) *Thought Without Language*. Oxford: Clarendon.

Libet, B. (1985) "Unconscious Cerebral Initiative and the Role of Conscious Will in Voluntary Action." *The Behavioral and Brain Sciences*, 8, 529–566.

Lockard, R.B. (1971) "Reflections on the Fall of Comparative Psychology: Is There A Message For Us All?" *American Psychologist*, 26, 168–179.

Mackintosh, N.J. (1985) "Contextual Specificity or State Dependent Learning of Human and Animal Learning." In L. Nilsson and T. Archer (Eds.) *Perspectives on Learning and Memory*. Hillsdale: Lawrence Erlbaum Associates.

Marcel, A. (1983) "Conscious and Unconscious Perception: Experiments on Visual Masking and Word Recognition." *Cognitive Psychology*, 15, 197–237.

Menzel, E.W. (1978) "Cognitive Mapping in Chimpanzees." In S.Hulse, H.Fowler & K. Honig (Eds.) *Cognitive Processes in Animal Behavior*. Hillsdale: Lawrence Erlbaum Associates.

Neisser, U. (1976) *Cognition and Rlity*, Freeman: San Francisco.

Povinelli, D.J. (1987) "Monkeys, Apes, Mirrors and Minds: The Evolution of Self-Awareness in Primates." *Human Evolution*, 2, 493–509.

Premack, D. and G. Woodruff (1978) "Does The Chimpanzee Have A Theory of Mind?" *The Behavioral and Brain Sciences*, 1, 515–526.

Roberts, W.A. and D.S. Mazmanian (1988) "Concept Learning at Different Levels of Abstraction By Pigeons, Monkeys and People." *Journal of Experimental Psychology: Animal Behavior Processes*, 14, 247–260.

Schachter, D.L., M.P. McAndrews and M. Moscovitch (1988) "Access to Consciousness: Dissociations Between Implicit and Explicit Knowledge in Neuropsychological Syndrome." In L. Weiskrantz (ed.) *Thought Without Language*. Oxford: Clarendon Press.

Schrier, A.M. (1985) "Perceptual Context Effects on Line Discriminations By Monkeys as Measured By Choice Response Accuracy and Eye Movements." *Animal Learning and Behavior*, 13, 392–396.

Seligman, M.E.P. (1975) *Helplessness: on Depression, Development and Death*, San Francisco: Freeman.

Sherry, D.F. (1988) "Learning and Adaptation in Food-Storing Birds." In R.C. Bolles and M.D. Beecher (Eds.) *Evolution and Learning*. Hillsdale: Lawrence Erlbaum Associates.

Squire, L. (1987) *Memory and Brain*, N.York: Oxford University Press.

Suzuki, S., G. Augerinos and A.H. Black (1980) "Stimulus Control of Spatial Behavior on The Eight-Arm Maze in Rats." *Learning and Motivation*, 11, 1–18.

Whiten, A. and R.W. Byrne (1988) "Tactical Deception in Primates." *Behavioral and Brain Sciences*, 11, 233–273.

Woodruff, G. and D. Premack (1979) "Intentional Communication in The Chimpanzee: The Development of Deception." *Cognition*, 7, 333–362.

Wright, A., H.C. Santiago, S.F. Sands and P.J. Urcuioli (1984) "Pigeon and Monkey Serial Probe Recognition: Acquisition, Strategies and Serial Position Effects." In H.L. Roitblat, T.G. Bever & H.S. Terrace (Eds.) *Animal Cognition*. Hillsdale: Lawrence Erlbaum Associates.

Wright, A.A., H.C. Santiago, S.F. Sands, D.F. Kendrick, and R.G. Cook (1985) "Memory Processing of Serial Lists by Pigeons, Monkeys and People." *Science*, 229, 228–229.

CHAPTER 2

USER MODELLING IN KNOWLEDGE-BASED SYSTEMS

M. Felisa Verdejo
Universidad Politécnica de Cataluña, Spain

I. INTRODUCTION

Computational models for interactive communication have been a growing area of activity in the last decade, both in theoretical investigation and as in practical implementations. Within this framework, we will pay attention to situations where different agents collaborate to solve a problem. In particular, we will now focus on the simplest form, considering two participants; a computer system and a user. Let us suppose they can communicate by formal, (restricted) natural language, graphics or any other means in order to achieve a given goal.

A major problem for cooperative communication is providing computer systems with capability to adapt themselves to each user so that on the one hand, they can offer support adapted to the user's competence and, on the other hand, they can inform accurately about their own knowledge and problem solving ability.

The purpose of this paper is twofold; first describe the type of necessary knowledge about a user in order to enable a computer system to carry out an effective communication. Second, to present different approaches to acquire and formalize this knowledge so that it can be handled in an automatic way.

In the second section some interactive situations are overviewed. A typologie is proposed for them in terms of the role played by the user in the communication process. Next, in section three, the content of a user model and some classification criteria are discussed. After that, two components of the model are studied; domain knowledge in section five, goals and plans in section six. Techniques to build models of different complexity are explored in more detail in section seven and eight. Finally, current results and guidelines for future research are summarized in the conclusion.

II. SITUATIONS OF INTERACTIVE COMMUNICATIONS

Question-answering systems are one of the fields where cooperative communication was first considered. Data Base Systems are a typical example. In this case the scenario consists of a user asking for information about the particular domain described in the data-base. The user formulates his requirements in a formal consulting language.

Then, the system processes the question and outputs either an affirmative answer consisting in the enumerations of the elements in the data-base verifying the properties required, or a negative statement.

This kind of interaction is based on several assumptions; first, the user knows what can be asked to the data base, and , therefore, all his questions are expected to be answered; second, the user knows how to ask questions about what he wants to know; and third, he should be able to interpret the answers correctly, specially the negative ones. In practice, the previous hypothesis demands that the potential users have a strong background on the database structure and content and therefore queries are usually formulated by data-base technicians.

According to Weber [Web-86], a question is a request to the interlocutor so that he provides the desired information and carries out the necessary actions to guarantee an effective communication. These actions will sometime consist in conveying supplementary information, as in example 1.

Example 1
Q- Where is the key to the secretary's office?
A- In the little cupboard, but it's closed.

Others, such as the one in example 2, offer a correcting answer which explains why the question (the way it has been formulated) has not had the expected answer.

Example 2
Q- Which is today's index?
A- Today, the stock market session in Barcelona has not finished, yet.

To provide a useful answer, it is necessary to be able to reason about the user's beliefs and intentions. In other words, we need to know the purpose of the question and the user's background knowledge. Therefore, to exhibit a cooperative behaviour, the system must include a user's model that can be exploited to elaborate a tailored answer.

Let us give another example to illustrate these ideas. Suppose somebody in a computer department asks the following question:

Example 3
Q- Where is the key to the terminals room?

Assuming (1) the final interlocutor's purpose is to enter the room to use the computer, (2) the computer does not work today, (3) the interlocutor does not know this fact, the right answer could be:

A- Today, the computer does not work.

The necessity of having and using a model of the user becomes even more clear when the complexity of the computer system increases. Advisory systems providing directives on how to develop a given task are a well known example.

Figure 1 shows an excerpt of a dialogue between a user and UNIX Consultant [Wil–88]. This system provides information on the commands to be given to the Unix operative system in order to obtain information, carry out different operations such as creating and consulting files, program execution, etc.

Each fragment corresponds to a user question and the answer offered by the help system respectively.

In fragment 1 we can see that, besides answering the question, the system warns the user on the necessary conditions for the command to work.

In fragment 2 the system informs that the required information cannot be provided, but indicates the user a way to obtain it by himself.

Fragment 3 shows a case of a negative answer, the system conveys supplementary information about the two topics appearing on the question: the command "rn"and how to rename. An example to illustrate the right use of the command is also given.

In fragment 4 the question is again formulated on an incorrect assumption and the system offers an informative answer indicating the non-existence of that option.

In cases where the computer system must interact with a variety of users, it is crucial not only to take into account user's goals, but also to give them an explanation adjusted to their expertness level, as illustrated in fragment 5.

Fragment 5 shows a question and two possible answers; the first is the one which would be given to a beginner, while the second would be provided to someone familiar with the system.

-Fragment 1-

How can one output John's files by the printer?
Using 1pr. But it does not work if you are not allowed to read John's files.

-Fragment 2-
Who is in the system now?
I am sorry but I can't answer this question. You can use the command users.

-Fragment 3-
Is rn used to rename a file?
No, rn is used to read news.
To rename a file you should use mv.
For instance, to rename file foo in the directory fool, type "mv foo fool."

-Fragment 4-
¿What does ls-v do?
Ls does not have v option.

-Fragment 5-
¿What does rhwho do?
a) Rwho is used to list all users of the network, their log-in moment and the time when they are not operating.
b) It is like who, but giving all the network users.

-Figure 1-

Expert systems are computer programs able to offer advice ranging from financial planning to medical diagnosis. These systems emulate the reasoning processes of human experts in a particular restricted domain area. A key feature is that, when requested by the user, the system offers justification or explanations for suggesting a specific solution, usually visualizing the successive steps in the reasoning chain.

A few attempts have been made to customize these explanations for the individual user, an example of such a system is UMFE [Sle–85].

The UMFE system adapts itself to the user by taking into account his background knowledge. When required, explanations are given in such a way that the reasoning chain contains only terms currently known to the user, avoiding concepts the system feels the user might not understand. For instance, if the casual explanation for a question is the following chain:

t1=>t2=>t3=>t4=>t5=>t6

and UMFE concludes t2 and t5 are unfamiliar concepts to a particular user, the explanation will be:

t1=>t3=>t4=>t6

Such assumptions are acquired by the system in the course of the interaction with the user, trying to ask him directly as few questions as possible. UMFE reaches a conclusion by using knowledge from the domain for a target population. This knowledge consists of the concepts involved, their importance and degree of difficulty, as well as rules expressing concepts associations like the following one:

if user X knows A, then, he also knows B, C and D.

Figure 2 shows the extract of an explanation dialogue between a user and UMFE acting as a front-end for MYCIN, an expert system on blood infections diagnosis.

user:	*What is the connection between* CRYPTOCOCCUS *and* INFECTIOUS *process?*
UMFE:	In this context, do you understand the significance of MENINGITIS?
user:	*Yes*
UMFE:	and FUNGAL MENINGITIS?
user:	*Yes*
UMFE:	In this context, do you understand the significance of CRYPTOCOCCUS?
user:	*Yes*
UMFE:	the inference chain is as follows: INFECTIOUS PROCESS==>MENINGITIS===> FUNGAL MENINGITIS==>CRTOTICICCYS
user:	*What is the connection between* COCCIDIOIDES *and* INFECTIOUS *process?*
UMFE:	INFECTIOUS PROCESS==>MENINGITIS===> FUNGAL MENINGITIS COCCIDIOIDES

-Figure 2-

The last situation we are going to explore concerns intelligent tutoring area. In this context, system-user interaction should be deep, involving a large amount of reasoning about the current state of the student's knowledge and intentions in order to understand his learning process. The example in figure 3 shows an excerpt of a dialogue between a student (A) and CAPRA [Gar-89], an automatic tutor for novice programmers. It corresponds to an intermediate session during which the student learns how to specify problems in a formal way.

T 1.1	Last day you learned to specify elements verifying a property (this property was expressed by a formula)
T 1.2	Do you have any questions about it?
A 1	No, go on, but could you give me an exercise?
T 2.1	I was going to do that
T 2.2	Specify next problem: "Express in polar representation (module, argument) a given complex number (real part, imaginary part)"
A 2	I don't remember what polar representation is. Could you tell me?

T 3 Given a complex number:
X coordinate--------real part
Y coordinate--------imaginary part
to express it in a (module, argument) form, you must calculate module as:
SQRT(X**2+Y**2)-----module
and argument as:
ARCTNG(Y/X)--------argument

A 3 Must I use the form of a pair?

T 4 Yes, you must

A 4 Can I use the concatenation instruction CONS?

.....

-Figure 3-

It can be seen that the teaching is carried out through a series of explanations, questions, drill and practice, etc., in a mixed-initiative dialogue.

The tutor's pedagogic knowledge (on how to teach) is expressed in terms of instructional goals with associated potential generic plans and rules to select and create an instance plan for a specific situation.

These rules include conditions involving considerable knowledge about the learner. A simplified example follows:

If the current student objective is a request for a teaching object and the pending tutor objective is to introduce a concept then acknowledge,
 establish next concept C and detail-level explanation L with student model Develop "Introduce concept C with detail-level L."

III. THE CONTENT OF THE USER MODEL

The previous examples can be classified along a scale of growing complexity in the richness of the interaction and, therefore, in the knowledge involved to carry it out. In summary, we have the following kinds of situations, presenting an increasing degree of cooperative behaviour:

1. The user asks questions on the explicitly stored information. The system works under the assumption that all the questions are properly formulated and only provides direct answers, either affirmative or negative, to those questions. Data Base Systems are a typical example.
2. The user asks questions on the supposedly stored information. The system outputs informative answers using reasoning techniques to interpret the user's question, as well as some meta-knowledge about the data-base.
3. The system is a problem-solver for a particular domain. The user provides information as required by the system or asks for explanations about the way the system proceeds. Expert systems are an example of this kind of situation. It is certainly acknowledged that a user model would improve the interaction process, allowing the system to communicate with users at their own level, but this aspect has not been used in current expert systems and remains a research topic.
4. The user wants to solve a problem. The system follows his activity and provides help, explicitly or non-explicitly asked for by the user. The role of the system is to act as an advisor, suggesting the right tools and how to use them. Therefore, a user model including his intentions and knowledge is required so that the system can offer him effective help.
5. The user is engaged in a learning process supervised by the system. In this case it is desirable to have a student model containing as much information as possible: background knowledge, preferred learning style, aptitude and motivation, learning goal, etc.

We have identified diverse situations where information about the user plays a significant role. Usually, five classes of components are distinguished in modelling a user: his knowledge, intentions, capabilities, preferences and motivations.

Knowledge can be further classified into domain-knowledge, reflexive knowledge (beliefs about his own knowledge) and reciprocal knowledge (the one he believes the system knows). Depending on the kind of situation, a variety of user's model functionalities is required and, according to this, the model content, its scope and granularity level also differs.

The principal aim of a user model is to suppose the system in adapting its behaviour, rather than to force users to conform to the same communication patterns. Two related aspects merit some comments; first, to remark that a user's model makes sense only if the system can adapt its behaviour. Under any other circumstance, it would be useless to discuss the existence of such a model. Lastly, to point out that the expressive power of a model is conditioned by the methods to acquire it. In general, the richer the model, the more complex the techniques to be used for its elaboration. Therefore, the final solution, in each case, must find a right balance between functionalities required and model construction complexity.

Before going into details about representation techniques for the various model components mentioned above, let us examine some dimensions used for model classification.

IV. CHARACTERISTIC DIMENSIONS OF A USER'S MODEL

First of all in order to be of interest, a user model has to be explicit, it means that information about the user must be explicitly encoded and grouped in a separate module. Furthermore, the notation used should be a knowledge representation language, i.e., rich enough to provide inference mechanisms. There are several criteria, some of them interdependent, to take into account in defining a concrete user's model. These criteria can be seen as different dimensions, in such a way that a final model is specified by a set of features resulting from choosing different options along this kind of multidimensional space.

The first dimension is related to the use of the model. A **descriptive model** encodes those aspects of the user which are going to be modelled, in a non-executable form. Attributes can have quantitative or qualitative values. The model is inspectable and provides support for user classification. Nonetheless, it cannot be used to reproduce the user's behaviour, while **predictive models** are in form of executable versions. This allows the system to perform user simulation and, hence, provides ways to anticipate his behaviour.

The second dimension concerns the extent of the user's domain-knowledge to be represented. **Overlays** models consider the user compared to the stereotypical expert, and explicitly represent the user's knowledge as a sub-set of an expert. Conversely, **differential models** consider only the variations on key issues between the user and the system.

Another alternative is the **genetic graph** proposing to make explicit the evolutionary relations between pieces of knowledge and including "deviation" links to represent also the user's misconceptions.

Finally, the third dimension establishes the alternative between surface model and deep model. The former aims at optimizing the performance, and then represent expertise in the most effective way without paying attention to its expressive adequacy, while the latter organizes knowledge following explicit epistemic criteria supporting behaviour explanation. It has been pointed out that one of the significant differences between experts and novices is the way in which they articulate their knowledge: compiled versus decompiled form respectively. However, it has also been observed that experts facing novel situations or justifying their behaviour are able to change from one representation to the other. Both approaches seem useful and multiple models can be considered although how to map from one to another is still an open question.

Three other dimensions are significant, but being related to model construction we will examine them later. Next, we are going to focus on two aspects: domain-knowledge modelling and intention modelling.

V. DOMAIN-KNOWLEDGE: SHALLOW VERSUS DEEP MODELLING

We can consider knowledge user compiled form when it has been especialized for a concrete use, loosing in that way generality and transparency. There can be several motivations. For instance, to configurate "automatic" behaviours in specific circumstances or for efficiency reasons when developing activities not requiring different levels of abstraction. Nonetheless, when on looks from the perspective of knowledge communication, that is the explanation or justification of problem solving behaviour, the need to have an articulated model of the domain has been proved by many experiments.

According to Wenger [Wen–87], knowledge articulation can be carried out with regard to two dimensions: the content of the model and its organization. With respect to the content, we can consider two aspects:

the first one is to reify justifications to be able to reason about, for instance what gives support to a piece of knowledge. This can be done by including a meta-level in the representation.

The second one would be the analysis of the forms in which new knowledge can be integrated in the model, as for example by transference, generalization, differentiation,....Regarding the organization, two criteria are significant: how to divide knowledge into units and how to give shape to these units.

In other words, we need to define epistemic categories to represent either conceptual or functional constituents. This research line is currently pursued in two fields: explanation and knowledge acquisition.

As Clancey points out in a well-known paper [Clan–84], most nowadays expert-systems represent knowledge in a form adequate to emulate expert performance efficiently. However, these knowledge bases cannot provide deep explanations or justifications about how problem-solving is being achieved. Opposed to the classical view characterizing reasoning according to the type of rule chaining mechanism implemented in the inference engine, Clancey proposes to distinguish three components: (1) general model, where, among others, common sense knowledge, taxonomic knowledge and heuristic associations are included. (2) generic tasks involved, for which he has established a typology, such as heuristic classification, aggregation,...and (3) specific problem model.

Heuristic classification is a generic task aiming at identifying a solution, among a known and pre-established set, which best solves a given problem. Pre-enumerated solutions are described by a set of abstract features and are organized in a taxonomic structure according to different criteria. This paradigm has been applied for medical diagnosis, troubleshooting, etc.,...inseveral expert systems.

Heuristic classification is mainly decomposed in three subtasks, as illustrated in the diagram below; data abstraction, heuristic matching, and solution refinement.

```
                         HEURISTIC MATCH

Data abstraction================>Solution abstraction
  |                                     |
ABSTRACTION                         REFINEMENT
  |                                     |
  Data                              Solutions
```

-Figure 4-

There is a line going upwards, from data to an abstract representation that involves inferences including taxonomic domain knowledge such as definitions and generalizations. Horizontal movements use operative knowledge, in form of heuristic associations, to match abstract data with abstract solution classes.

The other vertical line connects downwards abstract solutions to concrete ones by successive refinements. Each line represents a chain of several steps.

An example from MYCIN system is given in figure 5.

```
                         HEURISTIC RULE
Compromised host =============>Bacteria normally found
   |                            causing infection
GENERALIZATION                         |
   |                            gramnegative inf.
Immunosuppresion                       |
   |                            E.Coli, pseudomonas,
GENERALIZATION                  klesbiella
   |
  Leukopenia
   |
DEFINITION
   |
  Low WBC
   |
QUALITATIVE ABSTRACTION
   |
  WBC<2500
```

-Figure 5-

These ideas guided the design of Heracles [Cla–86]; a generic system which can be applied for building expert-systems performing heuristic classification. A typical rule looks as follows:

> "If a disorder causes and effect and there is evidence that the effect is present in the system being diagnosed, then there is evidence that the disorder is present in the system."

```
(SYSTEM patient)
   (DISORDER infection) (EFFECT fever)
   (CAUSES infection fever)
   (TASK: TEST-HYPOTHESIS)
   (FOCUS: (DISORDER X))
   If (AND ((CAUSE XY)(PRESENT YS))
   then (PRESENT XS)
```

It can be seen that the explication of categories such as system, disorder, effect...as well as the causal relationship between trouble and effect, permits to state domain terms and relations in a more abstract and functional way, using the diagnosis general model.

In the same way, task and focus explicitly state information related to the problem-solving strategy. Note that the level of abstraction in which rules are expressed is domain-independent and linked to the generic task performed. For the sake of comparison, the same rule in Mycin is expressed as:

```
IF (AND (PATIENT X)(FEVER X) then (INFECTIONS X))
```

If the patient has fever then he has an infection.

It is currently acknowledged that, in order to provide explanations, deep models like the one discussed above are needed for domain representation. A further point is to what extent user domain model has to conform also to this framework.

VI. MODELLING INTENTIONS

There are two basic approaches to determine user's goals; one, we can make the user express them in an explicit way; two, we can try to infer them from observing user behaviour. Systems providing natural language communication are usually based on the second approach, while those following the first option offer interactions based on menus tailored to the domain-problem.

The relation between language and action, that is, the recognition of the interlocutor's intention by interpreting what he says, has been widely studied in different fields. In Computational Linguistics, speech act theory [Sea–69] has received the most attention, and planning techniques are the approach proposed to implement a computational model of communication.

A plan is defined as a sequence of actions which, when carried out, enable us to achieve a goal. Actions can be modelled as functions from one world state to a new state that is the resulting world after the action has occurred. Usually you define these functions by specifying preconditions, a set of formulas that must be true for the action to take place, and effects. The list of effects specify how to construct the resulting world.

For example, the action of a person going out of a room would be expressed:

Going-out (person,room)
Preconditions: The person is in the room, the person wants to go out of the room, the door of the room is open
Postconditions: The person is out of the room

In a similar way, speech acts can be defined. An instance of a rough definition for a request follows:

Request (speaker, hearer, action)
Preconditions: the speaker wants the hearer to perform action
Postconditions: the hearer wants to perform action

This action states that one way to make another agent want to do something is to request him to do it.

Suppose we want somebody to go out of the room. A possible plan is the following:

1. Request (Maria, George, going-out (George, room 1))
which has the effect
2. wants (George, going-out (George, room 1))
which is a precondition of the action going-out, assuming the other preconditions are true
3. Going-out (George, room 1)
with the desired effect
George is out of room 1

Automatic planning consists in building a sequence of actions to achieve a goal state starting from an initial one. A common strategy is backwards chaining reasoning: you start looking at the goal state, selecting a fact that does not hold in the initial state, and then select an action having this fact as one of its effects. If the preconditions of this action hold in the current state, the action can be applied and the difference is solved. In cases where preconditions do not hold, they become in turn subgoals and, through recursive reasoning until reaching the initial state, an action sequence can be determined.

For instance, let us suppose that the door is closed, then one of the needed preconditions for going-out does not hold, and this new subgoal is attempted. If there exists an action Open (person,door) whose effects include "the door is open," without preconditions, then the action is included as step 2. At that point, conditions for Going-out hold, obtaining in this case a plan formed by three actions. This approach can be viewed as a search process, and in fact it suffers combinatory explosion problems due to the number of subgoals to be solved and the different choices to attempt.

Another proposal is to represent an action in terms of other (simpler) actions, instead of reasoning over the states, now we use action decomposition, by successive refinements, until a level where generated actions are primitive. This method is called hierarchical planning.

User's goal can be inferred applying planning techniques. In this case the model must include domain actions, and very likely the kinds of plans the user might have. From the information provided in the interaction the task of the system is to recognize/reconstruct what plan the user is following and then infer his goal.

Let us consider the dialogue in figure 3. In A3 the student asks for confirmation about the form to be used to express the solution.

Must I use the form of a pair?

The tutor interprets this request as a step in a plan to reach the goal "solve the proposed exercise." The next figure shows part of the recognised plan in which the chain from the request to the action "specify the output" can be explicitly built by hierarchical planning techniques.

```
                    Solve exercise

part of                         part of
 \                               \
Specify input                   Specify output
                                 \
                                part of
                                 \
                                Specify main objective
                                 \
                                enables
                                 \
                                Know how to specify

                                \effect

                                be informed

                                \effect

                                request to be informed
```

-Figure 6-

A cooperative answer is provided:
Yes, you must

Giving the information needed by the user to complete his plan.

Knowledge about user's plan may be used also to warn him about possible obstacles. For instance, suppose, on the scenario above, the next question:

Can I use the concatenation instruction CONS?
In this case the system could answer:
No, CONS is a programming instruction. You are now in the specification phase.
Conveying this supplementary information is a way to indicate that his plan is inappropriate.

Several Natural Language Comprehension Systems [All–82], [Kob–88], [Lit–84] explore these issues in detail.

VII. BUILDING A USER'S MODEL

As mentioned before, three other dimensions are of interest: generic versus individual models, permanent versus temporal models, and predefined versus inferred models.

Once the need for modelling some aspects of the interlocutor has been established, we can consider three possibilities. The first one is to assume that users's behaviour is uniform enough to consider an homogenous model for all of them. For instance, a question-answering system about train journeys can reason about the travellers' intentions assuming all possible plans that may be used for that domain.

The second possibility would be having classes of users, which are called profile models. Let us suppose a friendly interface to a word processor, we can consider three levels of familiarity: novice users, users who do not know this particular processor, but have worked with similar ones, and users mastering the system. The interface will classify the user in one of them. The degree of help and explanations detail will be modelled according to the user's profile.

The third possibility is to consider individual models, reflecting each one of the users interacting with the system. The richer the interaction, the more the system will benefit from a user model to tailor its behaviour. Learner's models for intelligent tutoring systems are an example.

On the other hand, we must also take into account whether the model is for carrying out a particular interaction, the information about the user being forgotten afterwards or whether the model is permanent, and should be saved for future interactions. An alternative is to consider both models, an extended short term model being summarized at the end of the session in order to update a long term model.

Finally, as pointed out in the previous section, there are two basic techniques to build a user's model: recognition and construction.

The first approach is mainly a classification process in order to determine the predefined model that best fits the current user.

By contrast, model construction aims at acquiring knowledge about a particular user from the interaction process. Two options have been considered. The first consists in acquiring the information in an explicit way, either through direct or indirect questions. The design of interactions, through which the information needed could be obtained from the user in a non-intrusive way, is currently a focus of active research. The second is based on observing the user to infer his current state of knowledge & intentions. This inferring approach relies on extensive domain knowledge bases and expensive reasoning techniques and has received the most attention in the field of intelligent tutoring systems focusing on learner modelling for diagnosis purposes. In the following section learner's models are explored in more detail.

VIII. LEARNERS MODEL

Learning can be considered an activity of evolutive exploration/acquisition of a space of knowledge states and, under this paradigm, the tutor's task would be helping the student to attain a desired goal state exploring this space in an optimum way.

In order for the tutor to generate the next instructional move, adapted to the current user, a learner model is needed. Thus, teaching requires an explicit, individual model of the student, with short and long term versions. A part of this model should include static attributes like aptitude, motivation, preferred learning style, etc...to support the selection of appropriate instructional strategies.

A key function for a learner model is diagnosis, and this requires an explicit, and preferably executable, representation of the learner's domain knowledge. Extended overlay to cope with misconceptions is the option most included in existing ITS systems. Usually, the approach to acquire the learner model is constructive, relying on extensive domain knowledge base and following the student's learning behaviour in detail to infer his knowledge state dynamically. This task has been proved to be a very difficult one, and some researchers have questioned the practical possibility of building these models. There are, nevertheless, two ways in which building learner models are attempted: enumerative and reconstructive theories.

Enumerative theories are based on a catalogue as exhaustive as possible to the bugs that a learner might exhibit. Depending on the domain, these bug libraries are in form of incorrect procedures [Bro–78], mal-rules [Sle–81], or buggy plans [Spo–85], [Urr–88]. This approach is both time intensive and difficult. A deep and detailed analysis of student errors is needed and this effort has to be done for each domain.

Bugs are usually represented in executable form to support analysis by simulation. Bug specification usually includes some meta-knowledge as, for example, source, frequency of occurrence, other related bugs, etc...as well as suggested remediation actions.

Diagnosis with bug catalogues ranges from search to heuristic classification.

Buggy [Bro–78] is a system for the diagnosis of errors in the domain of arithmetics. Some examples of the kind of errors included in Buggy's catalogue, corresponding to the subtraction procedure are the following:

a) The smallest digit is subtracted from the biggest one in each column, regardless of which is on the top
An example of this kind of bug:

```
 143
 -28
-----
 125
```

b) When the top digit of a column is 0, the bottom digit is the result (i.e. 0-N=N), an example:

```
 140
 -21
----
 121
```

The LMS system [Sle–81] performs diagnosis on solving algebraic equations. Figure 7 illustrates a fragment of its knowledge base; in A samples of correct rules are given, some mal-rules can be seen in B. C shows two resolution examples indicating the correct and buggy version for each case.

A. Some rules

rule name	level	condition	action
FIN2	1	(SHD x =M/N)	(SHD(MN)or(SHD evaluated)
SOLVE	2	(SHD M.X=N)	(SHD X=N/M)or(SHD INFINITY)
ADDSUB	3	(lhs M + I-N rhs)	(lhs [evaluated] rhs)
MULT	4	(lhs M*N rhs)	(lhs [evaluated] rhs)
XADDSUB	5	(lhs M*X+I-N*X rhs)	(lhs(M+I-N)*X rhs)
NtoRHS	6	(lhs+I-M=rhs)	(lhs+rhs-I+M)
REARRANGE	7	(lhs+I-M+I-N*Xrhs)	(lhs+I-N*X+I-M rhs)
XtoLHS	8	(lhs= +I-M*X rhs)	(lhs-I+M*X=rhs)
BRA1	9	(lhs<N>rhs)	(lhs N rhs)
BRA2	10	(lhsM*<N*X+I-P>rhs)	(lhs M*N*X+I-M*P rhs)

M, N, y P are integers, lhs, rhs, etc., are patterns, SHD stands for the head of the chain Y.

B. Some mal-rules

rule name	level	condition	action
M.SOLVE	2	(SHD M*X=N)	(SHD X=M/N)or(SHD INFINITY)
M.NtoRHS	6	(lhs+I-M=rhs)	(lhs=rhs+I-M)
M2.NtoRHS	6	(lhs1+I-M lhs2=rhs)	(lhs1+I-lhs2 = rhs+I-M)
M.XtoLHS	8	(lhs= +I-M*X rhs)	(lhs +I-M*X = rhs)
M1.BRA	10	(lhs M*<N*X+I-P>rhs)	(lhs M*N*X+I-P rhs)
M2.BRA	10	(lHS M*<N*X+I-P>rhs)	(lhs M*N*X+I-M+I-P rhs)

C. **Pairs of correct (left) and buggy (right) solutions**

i) Solving 2x=3*4+5

	2X=3*4+5		2X=3*4+5
MULT	2X=12+5	ADDSUB	2X=3*9
ADDSUB	2X=12+5	MULT	2X=27
SOLVE	X=17/2	SOLVE	X=27/2
FIN2	(17/2)	FIN2	(27/2)

ii) Solving 2x+5+9

	2X+5=9		2X+5+9
NtoRHS	2X+9-5	M.NtoRHS	2X=9+5
ADDSUB	2X=4	ADDSUB	2X-14
SOLVE	X=2/1	SOLVE	X+7/1
FIN2	(2)	FIN2	(7)

-Figure 7-

Enumerative theories guarantee bug detection under the assumption that the bug library is complete. However, experiments with previous systems proved that it is not usually the case. Conversely, Reconstructive theories propose to generate the rule or plan causing the detected error from the student resolution process, instead of identifying it in a pre-defined catalogue. Current attempts following these guidelines [Sle–83], [Lan–84] are restricted to very narrow areas.

A main problem is the big size of the search space (to determine possible mal-rules). It is proposed to use heuristics to guide the search together with inductive techniques. However the problem with using a machine learning approach is that the generated mal-rules may not support student modelling in the sense that they could be unrepresentative of student errors.

CONCLUSION

It has been argued that for a knowledge-based system to communicate in a cooperative way user modelling is needed. It becomes more critical as the establishment of a rich interaction is decisive for the agents to carry out the required tasks. Yet, the definition and acquisition of a user's model is a complex problem and, therefore, the practical solution will be in each case a tradeoff of the benefit of having a user model, in function of the desired versatility for the system and the range and variety of the communication process, against the cost of implementing effective resources to acquire the model.

Investigation guidelines interlace with problems in other areas, not only in Artificial Intelligence but also in Psychology, Computational Linguistics, Epistemology and Pedagogy. This is due to the basic and fundamental subjects involved, concerning the nature of intelligence: knowledge, its articulation, acquisition and communication on the one hand, and reasoning about beliefs, motivations, intentions and attitudes on the other hand.

The definition of better communication computational models will benefit from the progress in the mentioned fields, specially from works on knowledge acquisition and machine learning, since it seems clear that models should be defined by combining explicit and implicit techniques, based on communication process.

Although we take into account the close relationships with other disciplines, the framework within which we are working in Artificial Intelligence presents a singularity which must be pointed out: we are trying to create communication environments between a person and a computer or between computers. These are new kinds of situations that deserve a deep study. Human behaviour when dealing with computers is quite different from the one observed in people interaction.

For all these reasons, future work will evolve on two main directions. The first one is the study and characterization of communication situations with full details on the models of the agents taking part in the process. The purpose is to define the functionalities a model has to provide for each class of situations. The second one is to further explore effective ways for building these models with both explicit and implicit approaches, combining the design of better intercommunication methods with improved acquisition and learning techniques.

REFERENCES

[All–82]
Allen, J., C. Perrault. Analyzing intention in utterances. *Artificial Intelligence* 15, pg. 143–178, 1982.

[Bra–85]
Brachman, R., H. Levesque. *Readings in Knowledge Representation.* Morgan Kaufmann, Los Altos California, 1985.

[Bro–78]
Brown, J., R. Burton. Diagnostic models for procedural bugs in basic mathematical skills. *Cognitive Science*, vol. 2, pg. 155–191. 1978.

[Cla–83]
Clancey, W.J. The epistemology of a rule-based expert system: a frame work for explanation. *Artificial Intelligence*, vol. 20, no. 3, pg. 215–252. 1983.

[Cla–86]
Clancey, W.J. From GUIDON to NEOMYCIN and HERACLES in twenty short lessons: ONR final report 1979–1985. *A I Magazine*, vol. 7, no. 3, pg. 40–60.

[Gar–89]
Garijo, F., M.F. Verdejo, A. Diaz, I. Fernadez, K. Sarasola. CAPRA: an intelligent system to teach novice programmers. In *Perspectives of Artificial Intelligence*. Campbell & Cuena edt. Vol. 2, Ellis Horwood International, 1989.

[Kob–88]
Kobsa, A., W. Wahlster. Special Issue on User modeling. *Computational Linguistics*, vol. 14, nº 3, Setiembre, 1988.

[Lan–84]
Langley, P., S. Ohlsson, S. Sage. Machine learning approach to student modelling. *Technical Report CMU-RI-TR-84-7.* Robotics Institute, Carnegie Mellon University, Pittsburgh, Pennsylvania, 1984.

[Lit–84]
Litman, D., J. Allen. A plan recognition model for clarification subdialogues. *Proc of the 10th Int. Conf. on Computational Linguistics.*

[Sea–69]
Searle, J. *Speech Acts: an essay in the Philosophy of Language.* Cambridge University Press, Cambridge U.K. 1969.

[Sle–81]
Sleeman, D., M.J. Smith. Modelling student's problem solving. *Artificial Intelligence*, vol. 16, pg. 171–188. 1981.

[Sle–83]
Sleeman, D. Inferring students models for intelligent computer-aided instruction. *Machine Learning* pg. 483–510. 1983.

[Sle–84]
Sleeman, D. A user-modelling front-end subsystem. *Int. Jrnl Man-Machine Studies*, vol. 23, nº 1, pg 71–88. 1984.

[Ste–82]
Stevens, A., A. Collins. Misconceptions in student's understanding. En *Intelligent Tutoring Systems*. Brown, Sleeman (edt) Academic Press, 1982.

[Spo–85]
Spohrer, J., E. Soloway, E. Pope. A goal/plan analysis of buggy Pascal programs. *Human-Computer Interaction*, vol. 1, nº2, pg. 163–207.

[Urr–88]
Urretavizcaya, M., M.F. Verdejo. Estudio de modelos de estudiante y sistemas debugger de programación: su diseño en CAPRA. *Report Facultad de Informática 6/88*. Barcelona, 1988.

[Web–86]
Weber, B. Questions, answers, and responses: interacting with Knowledge Base systems. En Brodie, M., (eds). *On Knowledge Base Systems*. Springer Verlag, Berlin, 1986.

[Wil–88]
Wilensky, R., D. Chin, M. Luria, J. Martin, J. Mayfield, and D. Wu. The Berkeley UMX Consultant Project. *Computational Linguistics*, vol. 14, nº. 4, 1988.

[Wen–87]
Wenger, E. Artificial intelligence and Tutoring Systems. *Computational and Cognitive approaches to the Communication of Knowledge.* Morgan Kaufmann, 1987.

CHAPTER 3

CHANGING BELIEFS RATIONALLY: SOME PUZZLES

Dorothy Edgington
Birkbeck College, University of London
England

I. BACKGROUND

The puzzles I shall discuss belong to the subject names (by Richard Jeffrey[1]) "Probability Kinematics," that is, the question of how probability judgments should change in the light of new information. But I start with a bit of the pre-history of this subject as I see it.

Deductive logic, or the study of the principles of conclusive reason, made enormous leaps forward at the end of last century with the work of Frege, giving us logic much as we know it today. It was to be expected, and was expected, that inductive logic, or the study of principles of non-conclusive reason, would follow suit—that a comparably rich set of principles would be discovered, telling you that if you have premisses of such-and-such form, they render such-and-such a conclusion probable to such-and-such a degree. But despite much effort, nothing very impressive, or very realistic, emerged. Or more accurately, what emerged was impressive in inverse proportion to the extent to which it was realistic.[2] The question why the programme of constructing an inductive logic met with so little success demands a philosophically illuminating answer: is it intrinsic to the nature of such reasoning that it defies systematization? Did we set off on the wrong path, looking for the wrong kind of analogies with deductive reason? Or might more effort yield what has so far eluded us? These questions are important for, *inter alia*, the prospect of constructing computer models of such reasoning. I do not address them fully here, but I shall now offer (without fully endorsing) one diagnosis of the trouble.

The picture presented by inductive logic is that any probability judgment stands in need of justification, by (1) premisses, which are not themselves probability judgments, and (2) principles of reasoning which get you from these premisses to your conclusion. Now if we think of a probability judgment as an expression of a partial degree of belief in a proposition, and it is just a fact (albeit an important one) about beliefs that they are not all-or-nothing things but come in degrees, then this demand can seem like a prejudice. Nobody thinks that all *beliefs* can be justified in terms of *other* beliefs, on pain of regress. Why think this of partial beliefs? Why treat them as second-class citizens among beliefs?

If we drop the implied asymmetry between total and partial belief regarding the need for justification—if we allow that our ultimate premisses include probability judgments—then (so this line of thought goes) we don't need inductive logic: we have the deductive theory of probability to tell us, given some probability judgments, which others we are committed to. Logic can't tell you what to believe. It is a requirement of consistency or coherence between beliefs. All we can demand in the name of logic, once we allow partial beliefs, is that they be consistent. And probability theory gives you the logic of partial belief.

Actually, there is a non-sequitur in the above line of reasoning. Let us grant that it is foundationalist dogma to insist that all probability judgments stand in need of justification in terms of premisses which are certainties. It is equally extreme, anti-foundationalist dogma, to insist that none can be so justified. Rejection of the whole-scale foundationalist programme does not have the consequence that there could be *no* interesting normative principles of the type traditionally sought. But I do not pursue that possibility here. More to my present purposes is this: what allows this second perspective to deliver interesting, powerful conclusions about the notions of evidence and non-conclusive reasoning is *not* merely consistency between beliefs, but substantive theses about how one's probability judgments should change under the impact of new information.

Crucial to the enterprise is the notion of a *conditional probability judgment*: the judgment that B is probable *given* A. On the one hand, that is simply the ratio of the probabilities of A&B and A:

$$P(B/A) = P(A\&B)/P(A)$$

On the other hand, a given judgment of conditional probability is held, plausibly but substantively, to dictate how the judger should change her degree of belief in B, should she learn that A is true. An example:

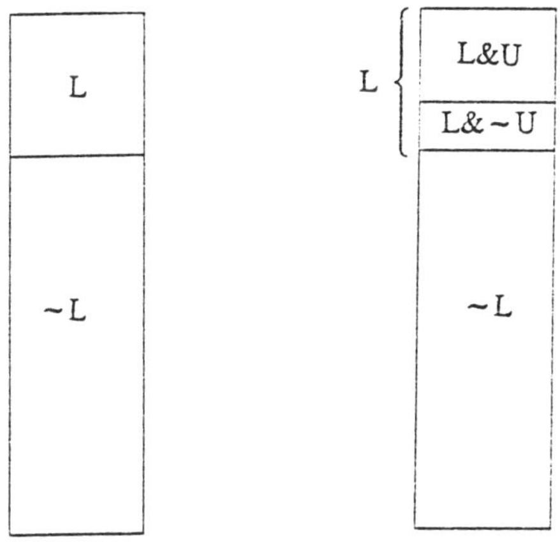

Figure 1

The above are notational variants of Venn Diagrams, with only one significant dimension, height. They are of height 1, and the horizontal lines indicate how a subject divides her belief between exclusive and exhaustive possibilities. An election is in the near future and L is the proposition that the Labour Party will win. On the left, we have a subject giving a lowish probability to this proposition. On the right, she has further subdivided this probability that the Labour Party will win into two, exclusive and exhaustive, parts, L&U and L&~U, where U is the proposition that unemployment will fall (by a definite amount in a definite time). Her *conditional* probability that unemployment will fall given that the Labour Party wins is best explained as follows. She supposes that the Labour Party will win; that is (in the right hand diagram), she supposes that the ~L-possibilities have been wiped out. Under that assumption—i.e., considering only the L-possibilities, the upper part of the diagram—

she considers how likely it is that U; that is, what proportion of the L-possibilities are also U-possibilities? This gives the standard equation:

$$P(U/L) = P(U\&L)/P(L)$$

Now suppose that the election results come in and she learns, to her surprise, that the Labour Party has won (and she learns nothing else of any relevance). What should she now think about P(U)? Conventional wisdom has it that she should now assign to U the probability she previously assigned to U conditional on L. This is *belief-change by simple conditionalisation*. Generalizing:

(i) On learning A, then, for any proposition B,
$P_{new}(B) = P_{old}(B/A)$

We shall look at the justification of this rule later. First we must generalize it, for it is unacceptably narrow in scope: it only deals with information which comes to us in the form of certainties. To hold that this is the only principle of belief revision would be, once more, too foundationalist: to hold that originating changes in belief are always of the form of unassailable facts. Beliefs acquired directly by observation are not always of that kind. Jeffrey raised this problem under the heading "observation by candle-light": I see a cloth in bad light and form a high degree of belief that it is green (but it may be blue). Examples are plentiful: I taste my dinner and say "I think it's veal—but it could be pork." I think he said "red"—but he might have said "led." In cases like this, there is, plausibly, no *proposition* describing the content of my experience of which I am certain: judgment or belief *enters* in uncertain form. Testimony from experts is another important source of new, but less-than-certain, probability judgments. What effect should new but uncertain information have on my other probability judgments?

Jeffrey's proposed rule I illustrate with another example of his, the Mud runner:

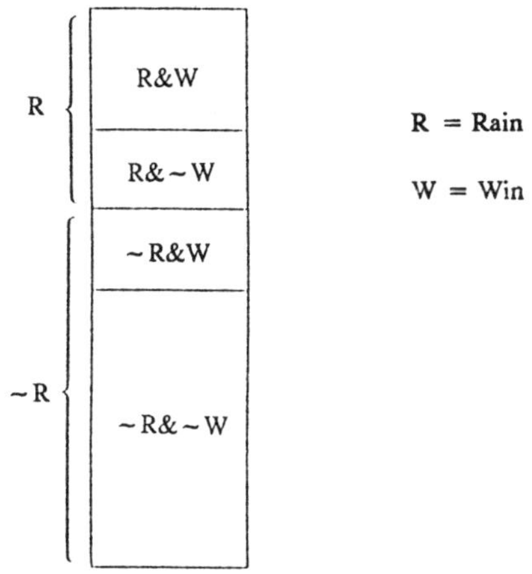

Figure 2

I am interested in this horse's chances of winning tomorrow. It runs relatively well on muddy ground: if there is rain, it has (I believe) about a 3/4 chance of winning; if it's dry, its chances are only about 1/6. Furthermore, I have it on authority that the chance of rain is about .4. This is enough to calculate that the horse's chances of winning are .4. Now I receive a new, updated weather forecast: the chance of rain is now judged to be .9. What effect should this new information have on my probability that the horse will win?

Jeffrey's answer is based on the following thought: changing one's degree of belief that it will rain should have no effect on the conditional probabilities, P(win given rain) and P(win given no rain). Thus, in the diagram, the effect of the new information is to increase P(R) to .9, and hence diminish P(~R) to .1, but *to leave unaltered the proportions of* R *and* ~R *which are also W*. Generalizing, this gives the rule: given a new value for P(A), then, for any B,

$$\text{(ii)}\ P_{new}(B) = P_{new}(A)P_{old}(B/A) + P_{new}(\sim A)P_{old}(B/\sim A)$$

This is called changing beliefs by *generalized conditionalisation* or *Jeffrey conditionalisation*. Simple conditionalisation is the special case where P(A) = 1.

II. A JUSTIFICATION OF GENERALIZED CONDITIONALISATION

These rules, as well as being very powerful, are intuitively appealing. Can anything more be said for them? I will give here a simplified version of a recent, little-known, and pretty argument of Bas van Fraassen's,[3] to the effect that if there is a rule determining one's posterior probabilities, given one's earlier probabilities and new information of this form, then this is the unique correct rule.

Van Fraassen's argument rests on two claims. First, if there is a rule of the above kind, then it must give the same answer for different problems of exactly the same structure; for that is just what it is for there to be a *rule* taking you from a prior distribution and a new value for one of the propositions in the distribution, to a posterior distribution. Thus, we might have the following problem structurally identical to the mud runner:

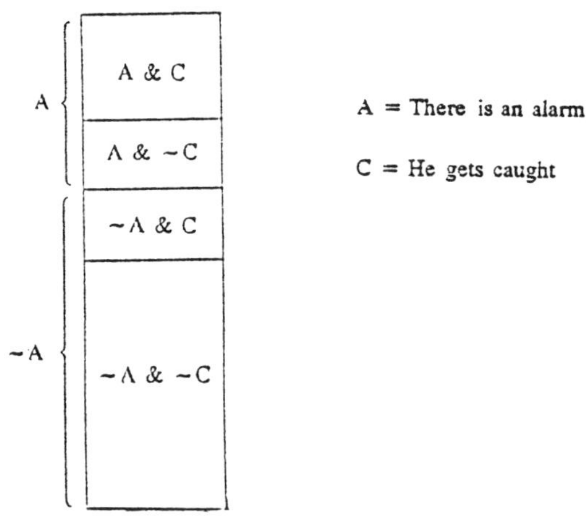

Figure 3

A burglar thinks his chances of being caught if there is a burglar alarm are 3/4; and his chances of being caught if there is no burglar alarm, 1/6. He thinks it is .4 that there is a burglar alarm. So he thinks it is .4 that he will be caught. Examining the wall in near darkness, he now sees something which makes him change his mind and think that it is .9 that there is a burglar alarm. What should be his new probability of being caught? Whatever it is, it must be the same as the new probability of the horse winning, given the structural identity of the two problems and the assumption that there is a rule. That is, there must be a structure-preserving mapping of the one final distribution onto the other.

Van Fraassen's second assumption is this: two people can have exactly the same probability distribution regarding the same subject matter, although one has it embedded in a finer distribution. That one has carved up the space of possibilities more finely, does not prevent them each, having ascribed the same probabilities to the same propositions, from facing the same revision problem subject to the same new information. Let us now see the mud-runner and the burglar problems stated in a grid of 10:

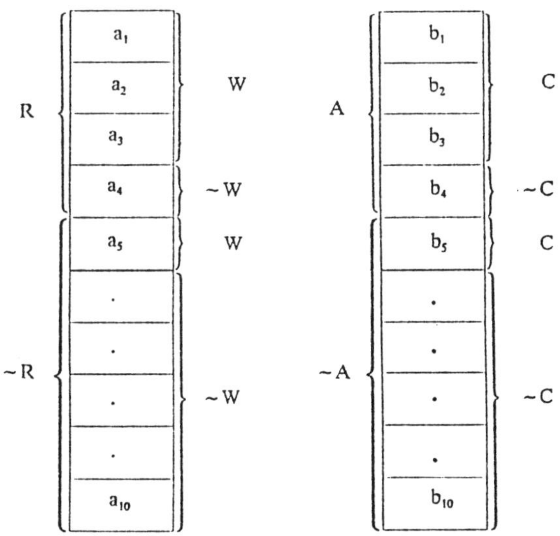

Figure 4

P(R) increasing to .9 in (1) has to have the same effect as P(A) increasing to .9 in (2), by the first claim. Therefore $P_{new}(a_1)$ must equal $P_{new}(b_1)$, $P_{new}(a_2) = P_{new}(b_2)$, etc....Now, all the relevant structure here is that P(R) consists of four equal parts, $a_1, \ldots, a_4$; and similarly, P(B) consists of four equal parts, $b_1, \ldots, b_4$. It is a matter of indifference which of the a_i we map onto which of the b_i ($i = 1, \ldots, 4$), when we claim that the new values in (1) and (2) must be the same. Thus $P_{new}(a_i) = P_{new}(b_i)$, for all $i = 1, \ldots, 4$. It follows by the transitivity of identity that all the $P_{new}(a_i)$'s are equal (e.g., $P_{new}(a_1) = P_{new}(b_1) = P_{new}(a_2)$, etc.). P_{new} (W) = the sum of the first three. Therefore, $P_{new}(W/R)$ remains unchanged at 3/4. QED.

This strategy will work for all rational values of probabilities. Some more advanced mathematics are needed to extend the result to the reals.[4] However, as assigning precise numbers to our degrees of belief is already an idealisation (though a fruitful one), the result for rational-valued probabilities is quite adequate for our needs.

This is a proof of uniqueness: if there is a rule, there is only one possible rule. What about existence? *Is there* a rule which determines what you should now believe, given what you believed before and your new information?

What we have proved is that, if nothing else is to affect your revised beliefs, other than your previous beliefs and your new information, generalized conditionalisation determines what you should now believe. Should anything else affect your new beliefs? If we interpret a negative answer strongly, as saying that any deviation from this dictate is irrational, it is unacceptable. New hypotheses can occur to you, new analogies between this problem and others, suggested to you by the new information, or otherwise, which may lead you to change your measure of the space of alternative hypotheses. Or you may just come to the conclusion that you placed too little weight on this factor, too much weight on that. But any deviation from the norm embodied in this rule needs a special story, a justification. (Indeed, if you do deviate from the norm, it seems likely that you will revise your view of what you ought to have believed *before* the new information, so that the effect of the new information still satisfies the rule.) The rule identifies the stable, fall-back position—it determines what you should now believe, given your new information and your old beliefs, and holding everything else constant. It properly represents the (direct) effect of the new information on your beliefs. Any deviation from it is a response to something else.

III. THE JUDY BENJAMIN PROBLEM

The above rule, (ii), though powerful, is still insufficiently general to cover all cases of belief revision with which we might be faced. For example, it is silent about (iii) how probability judgments should change in response to a new piece of *conditional* information, i.e., a revised judgment of conditional probability of the form $P(B/A) = r$. It is with this problem that my paper is chiefly concerned.

A general principle has been proposed, by, among others Peter Williams[5] which yields (i) and (ii) as special cases, and gives an answer to (iii) and much else besides. It is intended to be a principle of minimal change. We choose the posterior distribution of probabilities which satisfies the constraint imposed by the new information and contains the *minimum information relative to the prior*, (or alternatively, the *maximum entropy relative to the prior*) where the information in P relative to P^0 is given by:

$$I(P,P^0) = \Sigma p_j \log(p_j/p_j^0)$$

As a normative principle, this rule has, it seems to me, the disadvantage of being so complex that very few people could use it, or come to see it as compelling. (I would not make this complaint if it were put forward as a *descriptive* principle of how our beliefs *actually* change. A cat doesn't have to solve differential equations to be able to turn around in the air.) Although maximum entropy methods have some undoubted success in information theory, it is far from self-evident that they are compelling constraints on changes of belief. Bas van Fraassen suggested that the above formula gives the wrong answer to (iii) by means of an example inspired by the film *Private Benjamin.*[6]

During war games, Judy and her patrol are lost. They may be in their own Blue Army territory (B), or the enemy Red Army territory (R).

Each of these if further subdivided into Headquarters Company Area and Second Company Area (B_1, B_2, R_1, R_2). Assume that their initial probability-distribution is 1/4 for each of these four alternatives, and so P(B) = 1/2. They manage to establish radio-contact with their command, and after they describe their surroundings and what they can remember of their movements are told "I don't know if you're in Blue or Red, but given that you're in Red, the probability that you're in Red Headquarters is 3/4," i.e., "$P(R_1/R) = 3/4$." What should their final probability-distribution over the four alternatives be?

Everyone's intuitive answer, including van Fraassen's, is that P(B) and P(R) *should not be changed* by this new information, and $P(R_1)$ should be adjusted to be 3/4 of P(R), so the final distribution is 1/4, 1/4, 3/8, 1/8. For how can learning how likely it is that you're in R_1 *given* that you're in R affect the probability that you *are* in R? Maximum entropy, however, gives a different answer. P(B), $P(R_1)$ and $P(R_2)$ become 0.532, 0.351,0.117. Van Fraassen writes:

> The glaring feature of this solution is that the adjustment of Judy Benjamin's conditional probabilities in the Red Army Region have increased her subjective probability that she is still in friendly Blue Army Territory![7]

He adds that the effect becomes more extreme as $P(R_1/R)$ increases, and as the latter tends to 1, P(B) tends to 2/3. He concludes:

> It is hard not to speculate that the dangerous implications of being in the enemy's headquarters area are causing Judy Benjamin to indulge in wishful thinking, her indulgence becoming stronger as her conditional estimate of the danger increases.

In a final footnote added in press, however, van Fraassen changed his mind, in response to the following point put to him by Williams. Suppose the information had been that $P(R_1/R) = 1$. *This is just to be told that not-R_2!* So the effect should be the same as simple conditionalisation upon not-R_2, which yields that the four probabilities are now 1/3, 1/3, 1/3, 0, and P(B) = 2/3. So that P(B) tends to 2/3 as $P(R_1/R)$ tends to 1 is just what we should expect. And as we should expect the posterior distribution to vary continuously with $P(R_1/R)$, the maximum entropy result is not so surprising after all. Van Fraassen accepted this, abandoning the intuition that changing $P(R_1/R)$ should not affect P(B) and P(R); and his later work on the problem[8] takes as a constraint on a correct solution that P(B) tends to 2/3 as $P(R_1/R)$ tends to 1. Perhaps I am also engaging in wishful thinking: I like principles that I understand. I am going to try to defend van Fraassen's original objection to "maximum entropy"—that information about how likely it is that R_1 given R should not affect the probability of R. Van Fraassen reports (in the latest paper), and so do I, that this is (pretty well) everyone's initial reaction to the problem.

There is more at stake than this puzzle. The natural reaction is one which takes the conditional probability to be *epistemologically independent* of its "antecedent"—the proposition the probability is conditional upon—and that seems to me both correct and crucial to the dynamics of belief change. Consider the examples in Section I. Changing your mind about the probability of rain, it was argued, has, in itself, no effect on how likely it is that you think the horse will win *given* rain. This is the crucial principle behind simple conditionalisation and Jeffrey conditionalisation: changing one's degree of confidence in A has no effect on what you think about B given A. *And conversely*, I suggest. We should expect epistemological independence to be a symmetric relation, holding when there is no connection between one's grounds for two judgments. It is the fruitful juxtaposition of these two independently-grounded judgments which provides the basis for probability kinematics.

One often has firm beliefs about the probability of B given A, while having no opinion about how likely it is that A. For example, I think the probability is high that the match will light given that it's struck, while having no thoughts about how likely it is that it will be struck. There is one especially important context in which judgments of the form P(B/A) play a role when P(A) is undefined: decision taking. An agent is deliberating about which of a set of alternative courses of action, A_1, A_2, ...,A_n, to perform. Relevant to his decision is his judgment that $P(B/A_1)$ is high, $P(C/A_2)$, etc....Judgments about $P(A_1)$, $P(A_2)$, etc., are *essentially absent from his deliberation*. He has *not yet decided* which action to perform. (He may or may not be able to "look at himself from the outside," as a spectator, and judge the likely outcome of his deliberations. But that is to abandon the role of a deliberator.) This phenomenon does not invalidate the usual formula relating conditional probabilities to absolute ones. In judging P(B/A) to be high, I am committed to believing that *whatever* P(A) is (assuming it is non-zero), P(A&B) is nearly as much. But the phenomenon does show that the usual formula is not to be construed as a reductive definition of conditional probabilities in terms of absolute ones. A conditional probability has a life of its own, independent of the absolute probabilities whose ratio it is. The factors which determine our best estimate of it, then, should be independent of our estimate of the antecedent. If one must write "$=_{Df}$" somewhere in the usual formula, it would be more correct to write

$$P(A\&B) =_{Df} P(A).P(B/A)$$

recognising that P(B/A) is estimated independently, than to write

$$P(B/A) =_{Df} P(A\&B)/P(A)$$

(To have both as definitions would be circular.)

It is almost universally acknowledged that a conditional probability is a far better measure of the acceptability of a conditional thought than the probability of a material implication is.[9] What is wrong with the latter is the drastic lack of independence between A and ~AvB. Changing P(A) almost always changes P(~AvB), and, by the provable symmetry of probabilistic independence (see fn. 16), conversely. The classic objection to P(A⊃B) as a measure of belief in a conditional is "Just because it's now less likely that he will toss it (and hence more likely that either he won't toss it or it will land heads) it doesn't mean it's any more likely that if he tosses it, it will land heads!" In a sense, a conditional probability might be described as the probability of a material implication desensitised from its antecedent. If I believe ~(A&~B) *whatever I think about* A, then, and only then, do I judge P(B/A) to be high.

There are two possible strategies we could adopt to rebut Williams' argument against the intuitive thought that changing a conditional probability has, in itself, no effect on the antecedent. Williams made two claims:

> (i) when the conditional probability $P(R_1)$ changes to the extreme value, 1, then this is just to get the information that it is not the case that R_2. So the effect of this should be the same as simple conditionalisation on $\sim R_2$; this has the effect of making P(R), $P(B_1)$ and $P(B_2)$ each 1/3, and thus is inconsistent with the claim that P(R) is not changed.

> (ii) We should not expect a sharp discontinuity between the effect of $P(R_1/R)$ going to 1 and the effect if it goes to $(1-\in)$ for arbitrarily small $\in$. Indeed, we should expect arbitrarily small changes in $P(R_1/R)$ to have arbitrarily small effects on P(R). So for the non-extreme case, changing $P(R_1/R)$ will also affect P(R).

One strategy is to concede (i) to Williams, but to dispute (ii). I call this the weak strategy, because it concedes a special case in which change of conditional probability does affect the antecedent. The other strategy is to dispute (i), and to argue that there is a subtle difference between learning "You're not in R_2"—which does change the probability that you're in R, by simple conditionalisation; and learning "If you're in R, you're certainly in R_1," which does not change the probability that you're in R. I call this the strong strategy, because it holds on to the independence thesis unrestrictedly. In what follows, I attempt to defend the strong strategy. The arguments are less compelling than I would wish, and should they fail, I would retreat to the weak strategy. So first, I shall say something in favour of the weak strategy—why it would not be utterly counterintuitive and *ad hoc* to deny continuity, and treat the extreme value as a special case.

Firstly, in learning "If you're in R, you're certainly in R_1," you do learn a *fact*—a certainty. In all other cases you just acquire a new ratio your probabilities must satisfy. This is a qualitative difference, and it is perhaps not surprising that the case in which you learn a fact should have a different sort of effect on your beliefs.

Secondly, there are several principles of probabilistic reasoning which behave very differently for conditional probabilities of 1 and for conditional probabilities of $1-\in$, for any $\in$, however small. So there are sorts of precedent for the lack of continuity. I shall mention three such cases:

(a) Transitivity for conditional probabilities: assuming that none of the 'antecedents' have zero probability, if $P(B/A)$ - 1 and $P(C/B) = 1$, it follows that $P(C/A) = 1$; but we can have $P(B/A)$ and $P(C/B)$ arbitrarily close to 1, yet $P(C/A) = 0$. (Indeed, $P(B/A)$ can be strictly 1. As long as $P(C/B)$ is less than 1, $P(B\&\sim C)$ is greater than 0. Then let A be $B\&\sim C$.)

(b) Contraposition for conditional probabilities: if $P(B/A)$ is 1, $P(\sim A/\sim B) = 1$ (again assuming the 'antecedents' are non-zero). But $P(B/A)$ can be arbitrarily close to 1 while $P(\sim A/\sim B)$ is 0.

(c) Several philosophers have followed Ramsey in explaining acceptance of a universal proposition 'All Fs are G' as the disposition to believe, of any arbitrary thing, that it is G on acquiring the belief that it is F. If we interpret this 'acceptance' and 'belief' in terms of certainties, this is unexceptionable—but of little value as we hardly, if ever, do have such confidence in empirical universal propositions. On the other hand, I might assign a probability arbitrarily close to 1 to any arbitrary thing's being G, having discovered that it is F, yet be certain that *not* all Fs are G—when I know that there are many millions of Fs all but one of which is G, for instance.

So weak strategy is a viable fall-back. But I shall try to defend the strong strategy. Before doing so, we shall take a break from *Private Benjamin*. In the break, we shall look at two examples which (like the best commercials) are of some intrinsic interest, and may help to advertise how I believe she should solve her problem.

IV. AN APPARENT COUNTEREXAMPLE TO SIMPLE CONDITIONALISATION

If M is a master criminal, we'll never find out.[10] I believe that. Master criminals are extremely careful and skillful in covering their tracks. I believe that the consequent is probable given the antecedent; that is, that it is nearly as probable that: [she's a master criminal and we won't find out] as it is that she's a master criminal. But if I acquired the information that the antecedent is true, I would not assign a high probability to the consequent, as (apparently) I am required to do by simple conditionalisation.[11]

I shall argue that this is no counterexample. Conditionalisation does not require that *any* way of gaining the information that the antecedent is true obliges one to adjust the probability of the consequent to the previous conditional probability. I think it's probable that the match will light, given that it's struck. Other things being equal, information that it is being struck leads me to believe that it will light. But if I see you, match and matchbox in hand, dive to the bottom of the swimming pool and start to strike the match, *then* I don't think it will light. I *never did* assign a high probability to its lighting if struck in circumstances such as *these*. Of course, *I must have thought it unlikely that those circumstances would arise*, or else I wouldn't have thought it likely that it *would* light if struck.

In general, if I think it likely that B given A, I think that the A&~B circumstances are improbable relative to the A&B circumstances. But if I get the information that A by getting the information that an A&~B circumstance obtains, of course I won't believe B.

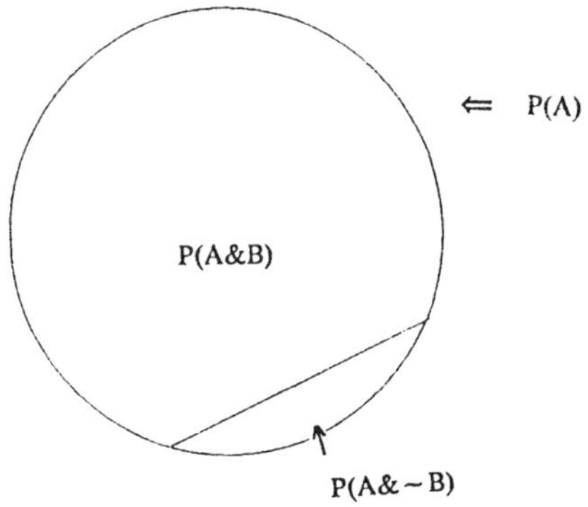

Figure 5

How, then, should we state what conditionalisation commits us to? Should we say, if P(B/A) is high and I get the information that A *and nothing else*, I should believe that B? We could, but this is difficult to apply. Information never comes one proposition at a time. If I learn by perception that you struck the match, I learn much more as well—the colour of your sweater, say, and the pattern on the wallpaper beyond. If I get information by testimony, I also get the information that the speaker believes it, and has decided to tell me, and standardly, much more collateral information besides. Still, A might be the only information I get *which is relevant to B*. The probability that the match will light, given that you strike it and are wearing a red sweater is the same as the probability that it lights, given that you strike it and are not wearing a red sweater. But the probability that it will light, given that you strike it and are at the bottom of a swimming pool is not the same as the probability that it will light, given that you strike it and are not at the bottom of a swimming pool. Conditionalisation, then, commits one to adjust P(B) to P(B/A) when A is the only relevant information obtained. In the swimming-pool case, I have additional relevant information.

Now, when I get the information that A, another piece of information which comes along free is the information *that I have the information that A*. This is a different piece of information. One is about the world, the other about me. One may be true, the other false. Usually, this is in the irrelevant, red sweater class. The probability that the match will light, given that it is struck and I have the information that it is struck is the same (in my opinion) as the probability that it will light, given that it is struck and I lack the information that it is struck. I do not believe that my information state affects the likely behaviour of matches. But in our conditional about M, the information, additional to the antecedent, that I have the information that the antecedent is true, could not be more relevant to the consequent. It is in the diving-to-the-bottom-of-the-swimming-pool class. And, just as conditionalisation doesn't commit me to believing the match will light when I get the information not only that it was struck, but was struck-in-the-improbable-way, so it doesn't commit me to believing I'll never find out when I find out not only that she is a master criminal, but that she's a master-criminal-in-the improbable (discovered) way. What is special about such conditionals with negative epistemic consequents is that one cannot discover that the antecedent is true without discovering that it is true-in-the-improbable-way —one always gains extra information relevant to the consequent which makes conditionalisation inappropriate.

This puzzle, then, only appears to be a counterexample to conditionalisation if we ignore the fact that as well as getting the information that A, I get further relevant information. The moral to be drawn is that care is needed about *precisely* what information one is responding to.[12]

V. THE THREE PRISONERS

Three prisoners, A, B and C, are told that two of them are to be hanged. A thinks "The probability that I will be hanged is 2/3." He asks the guard, who knows the fate of the individual prisoners "Will you give me the name of one prisoner, either B or C, who will be hanged?" The guard refuses to give any information. But A thinks, "The guard truthfully could have said 'B' or 'C.' If he had said 'B,' my chances would have been 1/2; and if he had said 'C,' my chances would have been 1/2. Either way, my chances are only 1/2 after all!" How can this thought-experiment reduce A's probability of being hanged from 2/3 to 1/2?

My first reaction to this puzzle was: the guard is irrelevant. The structure of the thought-experiment is: "Either B or C will be hanged. If it's B, my chances are 1/2; and if it's C, my chances are 1/2; so my chances are 1/2." The disjunction is known to be true. The conditionals, *interpreted as conditional probabilities*, are correct. But so interpreted, the consequents can't be detached. The premisses are consistent with A's chances being 2/3.

However, although these comments are correct about the reformulated thought-experiment, it is not equivalent to the first, and the interest of the puzzle as originally stated is deeper than that. Suppose the guard *had* responded to A's request, and said "B." What, then, should A judge the probability of his being hanged to be? Surprisingly, *the answer is not 1/2*, despite the fact that P(A/B) is, undoubtedly, 1/2 (writing "A" for A will be hanged, etc.).[13] Initially, there are three equally likely pairs, A&B, B&C, A&C. Suppose the guard says "B" in response to the above request, and assume he is truthful. P("B"/A&B) (i.e., the probability that he would say "B," given A&B will be hanged) is 1. P("B"/B&C) = 1/2. (He could just as easily have said "C" in this case.) And of course, P("B"/A&C) = 0. The guard was twice as likely to say "B" if A&B than he was if B&C. Applying Bayes' Theorem,[14] the correct posterior probabilities are P(A&B) = 2/3; P(B&C) = 1/3; P(A&C) = 0. A's chances remain, as they were before, 2/3.

(If you're not yet convinced, here is an analogous case: a fair spinner (disc with needle attached) is divided into three equal areas, numbered 1, 2 and 3. It also so happens that the disc is half red and half green: all of area 1 and half of area 2 is red, all of area 3 and half of area 2 is green. You will surely agree that the information that the needle landed on the red half makes it 2:1 that it landed in area 1. Now let areas 1, 2 and 3 represent A&B, B&C and A&C respectively, and let red represent the information "B." It could even be that the fate of the prisoners was decided by such a spinner, and the guard complied with the request for information by saying "B" if and only if the spinner landed in red.)

It is instructive to compare the manner in which A got the information that B would be hanged with another. Suppose A had asked the guard "Will B be hanged?" And the guard answered "Yes." This eliminates A&C but leaves A&B and B&C equally likely. In this case, A's probability changes to 1/2. Note that in this case, in asking the question, A was risking bad news: Had the guard said "No," A was certain to be hanged. He got good news by surviving the possibility of bad news. But with his original question, he was taking no risks. He already knew that either B or C would be hanged, and was merely asking for the name of one of them. Nothing could have been an unfavourable outcome, so nothing is a favourable outcome either.

Here again, we see how sensitive a posterior distribution is to *the manner in which information is acquired*. In a sense, in both cases, A just got the information that B would be hanged, but in one case he got it in a way that was twice as likely to "come from" A&B than from B&C; and in this case, his posterior probabilities are not got by simply conditionalising on B.

VI. JUDY BENJAMIN AGAIN: THE STRONG STRATEGY

Williams' claim that getting the information "*If* you're in the enemy region, you're certainly in their Headquarters area" is *just the same as* getting the information "You're certainly not in enemy Second company" must, I think, be resisted. Williams' claim is plausible. But in the second puzzle, it was easy to overlook the relevant difference between the information that A and the information that (A & I have the information that A). And in the third puzzle, it was plausible that getting the information "B" would have the same effect as the response to the two questions. So we should not be too surprised if there is a subtle difference between these cases too. If there is, we lose our motive for insisting that P(B) tends to 2/3 as $P(R_1/R)$ tends to 1, and are free to hang on to the original intuition.

Here is an example of the same structure: There are four horses in a race. Call them B_1, B_2, R_1, R_2. I am forced to bet, either that a B-horse will win or that an R-horse will win. I know nothing for or against any horse. I can make one quick phonecall to my expert friend, but I only have ten pence. Case 1: I tell my predicament. He says "R_2 won't win." The pips go and I am cut off. Case 2: The same except he says "(I don't know the B-horses but) if an R-horse wins, it will certainly be R_1." Now, it seems clear that in Case 1 my probability distribution changes from the uniform one to 1/3, 1/3, 1/3, 0, and I bet on a B-horse. It is surely less clear in Case 2. In case 1, the natural message is that R_2 is a bad horse. In Case 2, even if we eliminate the bracketed part of the remark, the message is that R_1 is better than R_2. It is as likely that R_1 is an outstanding horse, as it is that R_2 is a dud.

But here we come up against a complication which has been part of the Judy Benjamin problem all along: the Grice phenomenon. With freely offered information, we make inferences from the fact that the speaker said what he did as well as from what he said. When the context prevents further probing about the speaker's reasons for what he says, and it is important to glean the maximum from the brief exchange, the Grice factor is all the more prominent. In Case 1, the *natural message* is that R_2 is a dud, given that it was the first thing the expert said. But this is an inference not just from his speaking the truth, but from the assumption that he is being *helpful.* His truthfulness is compatible with his being sure that R_1 will win and that R_2 will come in second. But why should he have said "R_2 won't win," given that information and what he knew about my predicament?

Returning to Judy Benjamin, I want to argue that:

(1) Being told "You're certainly not in R_2"

is not equivalent to

(2) Being told "If you're in R, you're certainly in R_1"

And I want to guard against the danger that I am merely implicitly relying on the Gricean thought "Why should he have said (2) if he just wanted to convey what could be conveyed more briefly and less cryptically by (1)?"

If we are to factor out the Grice factor, the information we receive must be an answer to a question, one of a fixed range of possible answers. So let us ask: After describing her surroundings, etc., is

(1′) Asking the question "What (in your opinion) is the probability that we're in R_2?" and getting the answer "0"

equivalent to

(2′) Asking the question "Given that we're in R, what (in your opinion) is the probability that we're in R_1?"

(or equivalency: "What, in your opinion, is the ratio of the probability of R_1 to the probability of R?"

and getting the answer "1?"

It is not the same. For the answer to the second question could have been, in its second form: 0/0, or indeterminate; in its first form: undefined. In other words, (2′) gives the information, not only that she is not in R_2, but that, in the speaker's opinion, the probability that she is in

R is non-zero. (1′) does not give this information. And this difference is far from significant. Having described her surroundings, it was quite on the cards that the duty-officer would know she was in B, and give the "indeterminate" answer tot he second question. That he did not is a significant gain in information.

The background assumption of this story must be that the officer's superior knowledge of the terrain, etc., enables him to make better use of Judy's information about her surroundings than she can make herself, and that she should adjust her beliefs to her best estimate of his. (2′) is equivalent to (a) getting the answer "No" to "Is $P(R) = 0$?" and (b) getting the answer "0" to "What is $P(R_2)$?" (2′) is equivalent to (1′) only on the assumption that (a) should have no effect, and I cannot see why it should not. This is not a matter of Gricean implicature. (2′) entails $P(R) \neq 0$.[15]

VII. INDEPENDENCE

In this final section I give a valid argument from a plausible assumption to the conclusion that changing P(B/A) does not change P(A). I cannot claim that the conclusion is proved, because, unfortunately, the plausible assumption is false! However, I believe that the argument nevertheless lends credibility to the conclusion: the truth of the conclusion is the best explanation of some of the results which follow.

We tend to think of conditionals: judgments of the form "if A, B," as *propositions*. We may be uncertain about a conditional, just as we are uncertain about other propositions, but we think of this uncertainty as uncertainty about the *truth*—about the obtaining of some truth conditions, which identify the proposition in question (although, it is worth adding, no consensus has emerged among philosophers about what these truth conditions are).

Now it is hard to distinguish "It is probable that, if A, B" and "B is probable given A;" and there is much evidence that our degree of confidence in a conditional, "If A, B," goes hand in hand with our judgment of the conditional probability of B given A. So let us make an assumption: that there is a proposition, $A \rightarrow B$, whose probability is P(B/A):

(1) $P(A \rightarrow B) = P(B/A)$

We can prove, making essential use of assumption (1) that $A \rightarrow B$ is probabilistically independent of A; that is,

$$P((A \rightarrow B)/A) = P((A \rightarrow B)/{\sim}A) = P(A \rightarrow B)$$

and, probabilistic independence being probably symmetric,[16]

$$P(A/(A \rightarrow B)) = P(A/{\sim}(A \rightarrow B)) = P(A).$$

First we show that a conditional which satisfies (1) has the following two properties: (a) the truth of A& B is sufficient for the truth of $A \rightarrow B$; (b) the truth of A&~B is sufficient for the falsity of $A \rightarrow B$. (Indeed, on most theories of the conditional, these properties hold.)

> (a) Suppose A and B both have probability 1; then $P(A\&B) = 1$; $P(A\&B)/P(A) = 1$; i.e., $P(B/A) = 1$; so by (1), $P(A \rightarrow B) = 1$. Now suppose A&B were not sufficient for the truth of $(A \rightarrow B)$. It should follow that one could be sure that A and sure that B, yet not have enough information to decide whether $A \rightarrow B$; but this conflicts with the above; so A&B is sufficient for the truth of $A \rightarrow B$.

The argument for (b) is exactly parallel: suppose $P(A\&{\sim}B) = 1$. so $P(A\&B) = 0$; so $P(A\&B)/P(A)$, i.e., $P(B/A)$, i.e., (given (1)) $P(A \rightarrow B) = 0$. Now suppose A&~B were not sufficient for the falsity of $A \rightarrow B$. It should follow that one could be sure that A&~B yet not have enough information to decide whether $A \rightarrow B$, which conflicts with the above. So A&~B is sufficient for the falsity of $A \rightarrow B$.

Suppose A is true. Then, in the light of (a) and (b), B is true iff $A \rightarrow B$ is true; i.e., B and $A \rightarrow B$ are alike in truth value when A is true.

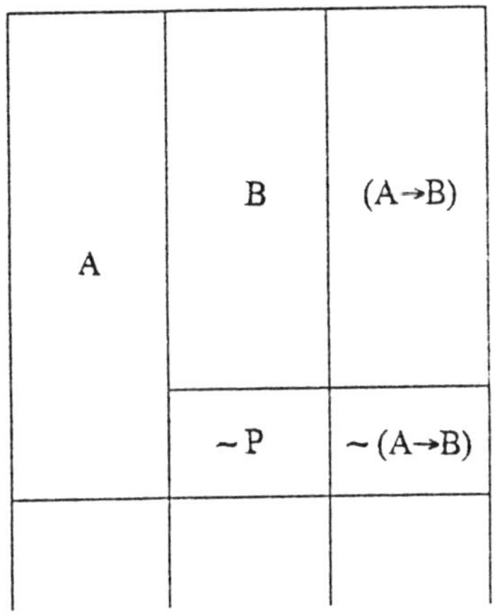

Figure 6

i.e., P(B/A) = P((A → B)/A). But, by (1), P(B/A) = P(A → B). Therefore, P(A → B) = P((A → B)/A); i.e., A → B is independent of A. With this result and assumption (1), we can show that new conditional information does not affect the probability of the antecedent: in the extreme case, to be told "The probability that you're in R_1, given that you're in R, is 1" is to be told $P(R \to R_1) = 1$. By simple conditionalisation,

$$P_{new}(R) = P_{old}(R/(R \to R_1)) = P_{old}(R)$$

(given the independence result).

In the non-extreme case, we had "The probability that you're in R_1 given that you're in R, is 3/4." By (1), we have $P(R \to R_1) = 3/4$. We can apply Jeffrey conditionalisation to derive

$$\begin{aligned} P_{new}(R) &= 3/4.\ P_{old}(R/(R \to R_1)) + 1/4.P_{old}(R/\sim(R \to R_1)) \\ &= 3/4.P_{old}(R) + 1/4.P_{old}(R) \text{ (by independence)} \\ &= P_{old}(R). \end{aligned}$$

It would be nice if that were the end of the story. But, unfortunately for present purposes, assumption (1) (though plausible) is false. As David Lewis proved, there is no proposition the probability of whose truth is necessarily equal to the ratio of the probabilities of A&B and of A.[17] The alleged conditional proposition in (1) behaves nicely enough in the circumstances in which A is true, but badly in the circumstances in which A is false. One example: suppose P(A) = 1/2, P(B/A) = 4/5 and P(B/~A) = 0. This is a perfectly possible distribution: see the two left-hand columns in Figure 7 below. We have seen that A → B must behave like B when A is true. And to make P(A → B) = P(B/A), we need P((A → B)/~A) = 4/5 also. We have now completed column 3 of the diagram. Note that P(~B&(A → B)) = 2/5; and P(~B) = 3/5. So P((A → B)/~B) = 2/3.

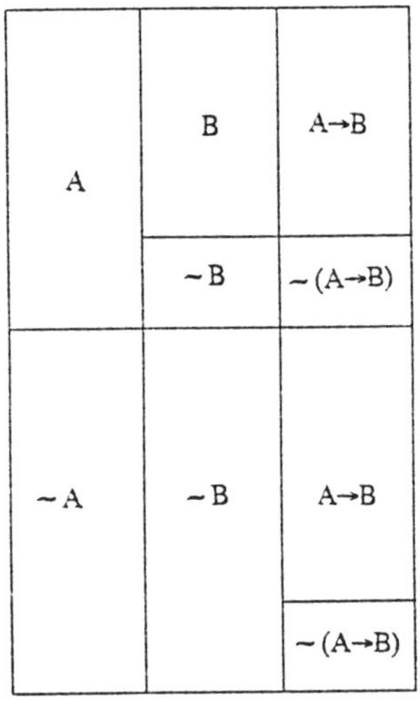

Figure 7

Now suppose I learn ~B. By the last formula and simple conditionalisation, I should judge P_{new}(A → B) to be 2/3. This is very strange: I know that B is false, I'm uncertain about A, but I think that it's

2/3 that if A is true B is true! As well as being strange, it is inconsistent with (1); for $P_{new}(B/A) = P_{new}(A\&B)/P_{new}(A) = 0$.

Note that this last is a perfectly reasonable result: Knowing that B is false, but being uncertain about A, I should surely give 0 credence to the conditional judgment that if A, B. The moral seems to be that conditional judgments *are* measured by conditional probabilities, but it is a mistake to think of conditionals as propositions, with truth conditions. I have argued this elsewhere.[18]

Some philosophers dispute the refutation of (1),[19] and others argue that it is of less significance than it seems, and a conditional may be defined which almost always, and in all but strange cases, satisfies (1).[20] For these philosophers, the above proof that changing conditional probabilities leaves antecedents unchanged should be convincing. I do not belong to this camp. I think the lesson of Lewis's proof is that conditionals should not be thought of as propositions, having truth conditions. But the result still seems to me to be significant: if there were a proposition whose probability was measured by a conditional probability, P(B/A), that proposition would have to be independent of A. The best explanation of this fact, surely, is that the judgment of conditional probability is itself something which neither affects nor is affected by the probability of its antecedent.

(A weaker conclusion is on even stronger ground: even if you don't accept the above as an argument for independence, you may agree that we have here an explanation of why everyone's intuitive judgment is that changing P(B/A) leaves P(A) unchanged; for P(B/A) measures what it is easy to confuse with the probability of a conditional *proposition*; and if it did measure the probability of such a proposition, that proposition would be probabilistically independent of its antecedent.)

NOTES

1. See Richard Jeffrey, *The Logic of Decision*, (McGraw-Hill, 1965) chapter 11.

2. I have in mind the work of R. Carnap, especially *The Logical Foundations of Probability*.

3. B. van Fraassen, "A Demonstration of the Jeffrey Conditionalization Rule," *Erkenntnis*, 24, 1986, pp. 17-24.

4. B. van Fraassen, *op. cit.*, pp. 22-23.

5. Peter Williams, "Bayesian conditionalisation and the Principle of Minimum Information," *British Journal for the Philosophy of Science*, 1980.

6. Bas C. van Fraassen, "A Problem for Relative Information Minimizers in Probability Kinematics," *British Journal for the Philosophy of Science*, 1981.

7. *Op. cit.*

8. R.I.G. Hughes and B.C. van Fraassen, "Symmetry Arguments in Probability Kinematics" in P. Kitcher and P. Asquith (eds), *PSA* 1984, vol 2 (East Lansing Michigan); and Bas C. van Fraassen, R.I.G. Hughes, and Gilbert Harman "A Problem for Relative Information Minimizers, Continued" (unpublished).

9. There is less agreement about how to explain this fact. David Lewis gives a Gricean explanation in "Probabilities of Conditionals and Conditional Probabilities," *Philosophical Review* 85, 1976, pp. 297-315, reprinted in his *Philosophical Papers* Volume II (Oxford, 1986). He later abandoned this in favour of Frank Jackson's explanation in "On Assertion and Indicative Conditionals," *Philosophical Review* 88 (1979), pp. 565-589, and "Conditionals and Possibilia," *Proceedings of the Aristotelian Society* 81 (1981) pp. 125-137. See also Ernest Adams, *The Logic of Conditionals*, (Reidel, 1975); and my "Do Conditionals Have Truth Conditions?," *Critica*, 1986, pp. 3-39, forthcoming in Frank Jackson (ed.) *Conditionals*, (Oxford: Oxford Readings in Philosophy, 1990).

10. This puzzle is a (depoliticised) version of one put by John Skorupski to Crispin Wright as a problem for an intuitionist-style conditional; and also, I heard, by David Lewis to Frank Jackson as a problem for Jackson's claim that conditionals are assertible only if they are "fit for modus ponens." *See* Frank Jackson, *op. cit.*. I have since discovered by solution to it is to be found, in essence, in Robert Stalnaker, *Inquiry*, (MIT Press, 1984), p. 105. See also David Lewis, *Philosophical Papers* vol. II, *op. cit*, pp. 154-156.

11. I believe, following Ernest Adams. *op. cit.*, that conditional judgments express high conditional probabilities. But you do not need to agree with me about that. Here, the puzzle is about high conditional probabilities, and conditionalisation.

12. My treatment of this example shows that it is not a counterexample to Jackson's claim: It is just that one can never get into the position to use *modus ponens* (one would if one could!), for one can never get the information that the antecedent is true without getting further relevant information. Whether the intuitionist-style conditional can be similarly saved is part of the general question of how to adapt logical constants designed for mathematics to empirical discourse. It is not my problem. See Crispin Wright, *Realism, Meaning and Truth*, (Blackwell, 1988), Appendix.

13. I am grateful to my ex-student Ruth Weintraub both for introducing me to this puzzle, and for being the first to suggest that A's probability of being hanged should not change in response to this information.

14. According to this theorem of the probability calculus,

$$\frac{P(H_1/E)}{P(H_2/E)} = \frac{P(H_1)P(E/H_1)}{P(H_2)P(E/H_2)}$$

15. Alex Edgington, my son, suggests the following (which is somewhat above my head). (It is framed in terms of a second question-and-answer equivalent to my 2′. "Given that we're in R, how likely is it that we're in R_2?" "0"): "another way of showing that the amount of information we get depends on how we ask the question is as follows: suppose Judy Benjamin assigns some initial probability density to the speaker's degrees of belief (which are themselves probabilities) in the four possibilities. From this density function, which is over four real variables, we can derive density functions corresponding to variables defined in terms of those four—in this case we are interested in "R_2" and "$R_2/(R_1+R_2)$." When we evaluate these derived densities at the answer received (0 in both cases), we get a measure of how plausible she would have thought that answer, and hence (inversely) of how much information it gives her. The point is that value is different for different variables ("R_2" and "R_2/R_1+R_2)"), even if the answers tell her the same fact about R_2. For example, suppose Judy assigns the following distribution to the speaker's beliefs: R_2 is distributed uniformly on [0,1] and R_1 uniformly on $[0,1\text{-}R_2]$. It is then easy to show that the p.d.f. of "R_2" evaluated at 0 is 1, whereas the p.d.f. of "$R_2/(R_1+R_2)$" evaluated at 0 is infinite (so she gets more information in the first case)."

16. P(A&B) = P(A)P(B/A) = P(B)P(A/B). So P(B/A) = P(B) iff P(A/B) = P(A).

17. "Probabilities of Conditionals and Conditional Probabilities," *op. cit.*

18. See my "Do Conditionals Have Truth Values?," *op. cit.*

19. See Simon Blackburn, in Charles Travis (ed.) *Meaning and Truth*, (Blackwell, 1986).

20. Bas van Fraassen, "Probabilities of Conditionals" in W.L. Harper and C.A. Hooker (eds) *Foundations of Probability Theory, Statistical Inference, and Statistical Theories of Science*, Volume 1 (Reidel, 1976), and Brian Ellis, *Rational Belief Systems* (Blackwell, 1979), pp. 74–80; and "Two Theories of Indicative Conditionals," *Australasian Journal of Philosophy*, 1984.

CHAPTER 4

ON THE REPRESENTATION OF LINGUISTIC INFORMATION

M. Teresa Espinal
Universitat Autònoma de Barcelona
Spain

I. INTRODUCTION

The aim of this paper is to discuss a series of problems raised in the representation of linguistic information or, to be more precise, in the representation of what has been called—from philosophical and linguistic approaches —parenthetical or disjunct constituents (cf., Urmson (1952) and Schreiber (1972)).

Specific reference will be made to some problems posed by the formal representation of the constituents underlined in (1) to (3).

(1) CATALAN

(a) La Maria, *almenys així m'ho han dit*, es casarà amb en Joan
(= Maria, at least this is what I've heard, will marry Joan)

(b) *Francament, noia*, no sé com sortir-me'n
(= Frankly, my dear, I don't know how to handle that)

(2) ENGLISH

(a) He is, *I believe*, a very powerful person
(= Es, creo yo, una persona con mucho poder)

(b) What is AIDS? *Quickly!*
(= ¿Qué es el SIDA? ¡Rápido!)

(3) SPANISH

(a) Ricardo, *que tiene una pequeña renta*, todavía está soltero
(= Ricardo, who lives on a small income, is still a bachelor)

(b) Se trata, *respondió su padre*, de un hombre que tiene el sentido del humor
(= He is, answered his father, a man who has a sense of humour

I shall attempt to show that the analyses proposed so far both from a philosophical point of view and grammatical theory do not provide an adequate linguistic description of this sort of data. It is particularly necessary to question the validity of the Theory of Speech Acts (proposed from the very beginning to solve philosophical questions derived from the notion of intentionality) when applied to the description of parenthetical units. If we seek to understand the way in which linguistic information is structured and the way in which human beings access the mental representation of certain linguistic sequences of their natural language, we must start from different assumptions.

I shall try to show that, in order to account for the interaction between disjunct constituents and syntactic structures and to explain what the interpretation (and, therefore, the mental representation) of an utterance containing a parenthetical constituent consists of, the modularity hypothesis (as it has been settled in linguistic studies) does not properly account for the data. This is because the information contained in those utterances is not encapsulated within a unique sentential structure and is not isolated from the central cognitive system, which means that parenthetical constituents are not syntactically integrated within the structure of the host sentence and, moreover, their processing is not controlled by strictly linguistic principles, but depends on principles —independently motivated—which seem to control the interface between different faculties of human cognition.

I am also concerned with the sort of information transmitted by an utterance containing a disjunct constituent and, consequently, I shall analyze some hypotheses which attempt to account for the way this information is structured and organized within oral communication. The present discussion agrees with the thesis defended by Pollard and Sag.

(4) The specification of a token linguistic object comes about in a cumulative (or *monotonic*) fashion via the interaction of constraints arising from several sources, including lexical

entries (which contain phonological, syntactic, and semantic information), the grammar rules that combine them, language-specific and language-universal principles of well-formedness, as well as the particular language-use situation itself (i.e., contextual factors); the final result is obtained by unifiyng the information from all these various sources (cf., Pollard and Sag (1987:7)).

Apart from this unification of the information, required to interpret complex utterances such as those in (1) to (3), the analysis of the data here displayed will allow us to argue for an extension of the theoretical apparatus originally postulated in autosegmental phonology, in such a way that the representation of disjunct constituents which appear in combination with sentences consists of a sort of multidimensional syntactic structures whose synchronization is dependent on certain restrictions and principles of a syntactic, conceptual and pragmatic nature.

Finally, I would like to point out that what will be said in the following sections will question the standard methodological partition between syntax, semantics and pragmatics, understood in the terms Hornstein (1986) has ironically referred to as: "Syntax the father, semantics the son, and pragmatics the ghost."

II. THE MODULARITY HYPOTHESIS

One of the central theses of the linguistic analysis developed within the Generative Grammar (GG) framework is the modularity hypothesis. According to it, one estimates that linguistics is part of cognitive psychology and that the grammar of a natural language is composed of a finite series of systems of principles and rules distinct from those which govern other cognitive systems.

(5) The modularity hypothesis can be regarded as the very strong psychological claim for the direct participation of the constructs of linguistic analysis in the process of language comprehension, up to and including the level of logical form. It saves a central role for syntactic analysis and representation, safely encapsulated within the language module (cf., Marslen-Wilson and Tyler (1987:61)).

In fact, perhaps it should be pointed out that the modularity of language has been mostly infered from the modularity of linguistic analysis, as postulated by philosophers such as Jerry Fodor and linguists such as Noam Chomsky.

According to Fodor (1983) a cognitive modular faculty is characterized by a set of features. The domain specificity refers to the fact that its operations are idiosyncratic and not found in other domains (i.e., its cognitive organization is vertical). Fodor also upholds that modular cognitive systems (such as the linguistic system) are genetically determined, associated with specific, localized, structured neural systems and computationally autonomous. Moreover, modular processes are said to be applied fast and mandatorily to a certain level of perceptual representation (i.e., the region of Logical Form; from now on LF).

Nevertheless, among all these features, the most important when referring to modularity is—according to Fodor—the encapsulation of information (also called the cognitive impenetrability of language perception): input or perceptual systems (which are modular) are isolated with respect to the information involved in central systems, for the information entertained in these systems is understood not to directly affect the processing of the former.

Against this hypothesis Marslen-Wilson and Tyler make the following observation:

(6) To defeat Fodor's argument, we do not need to show whether or not central and modular processes share the property of encapsulation, but simply that they are not isolated from each other in the ways the modularity hypothesis requires (cf., Marslen-Wilson and Tyler (1987:41)).

Setting up the question of exactly what degree of isolation the hypothesis requires, these authors notice that the parsing of certain syntactic sequences depends on extralinguistic information (i.e., encyclopaedic knowledge) and, therefore, that contextual information may affect the assignment of a syntactic analysis.

Similarly, I think that the processing and representation of linguistic sequences with parenthetical items pleads for a model that is more connectionist than modular. There are various reasons for this. First, because the disjunct information is not integrated within the hierarchy of constituents or the sentential domain of the host linguistic structure. Second, its processing occurs simultaneously with the processing of the sequence with sentential structure and, quite often, requires the existence of a specific connection with information (i.e., beliefs, presuppositions, pragmatic principles, etc.) which is outside the scope of the module of language, as it is conceived under the modularity hypothesis.

Consider briefly the example given in (7):

(7) Recordo la petita A.—*catorze anys—quan la tardor passada xuclava, amb els ulls tancats, un raïm moll i daurat*
(= I remember little A.—fourteen years of age—when last autumn she was sucking, with her eyes closed, a wet golden grape)

Observe that the two underlined constituents introduce additional assumptions of the speaker within the discourse. Furthermore, it is evident that those disjunct constituents are pieces of information syntactically and semantically dissociated from the proposition they have to be interpreted with at the very moment of utterance processing. It is due to factors similar to these that some cognitivists have objected to the modularity hypothesis. Look, for example, at (8), where Marslen-Wilson and Tyler express their divergence from Fodor.

(8) We reject [the modularity thesis] as a claim about the relationship between linguistic construct and psychological process. The facts of psycholinguistic performance simply do not suport the rigid dichotomy between the domains of the syntactic and the non-syntactic that is the central claim of the modularity thesis (cf., Marslen-Wilson and Tyler (1987:61)).

In other words, if our ultimate goal is the comprehension of language, we must account for the problem posed by linguistic sequences such as those in (7), for which it is possible to provide a correct syntactic analysis just in case adequate contextual information is available. Access to contextual assumptions such as *little A. was fourteen years of age* and *last autumn little A. was sucking, with her eyes closed, a wet golden grape* is a necessary piece of information in order to recognize that the whole syntactic sequence is not distributed under a single original sentential node, and in order to know that the disjunct constituent—in addition to not being an argument, a predicate, or an operator with respect to some other sentential constituent—cannot be substituted for any valence requirement of the verbal head of the sentence.

This means that an on-line interactive language processing theory, which presupposes an on-line availability of information from different levels of representation, comes into play. But, what is even more crucial is the fact that in examples such as (7) the meaning of the discernible units and context seem to influence the syntactic analysis adscribed to the linguistic entities it is composed of.

I do not think we really face an argument for an on-line weakly interactive theory of language processing, because according to this interpretation (cf., Carston (1988:52)) "syntactic processing independently sets up the alternatives which semantics and context choose amongst." What (7) presents is clearly not an example of postperceptual disambiguation but a case where extramodular information guides linguistic computation, and this is incompatible with encapsulation.

III. GRAMMAR, PRAGMATICS AND MODULARITY

In harmony with the modularity thesis, a complementarity of linguistic models and an autonomy of grammatical components have been proposed. More exactly, the linguistic community working in the GG framework has assumed an architecture of systems of linguistic principles which for every sentence of any language will relate its Phonetic Form with its LF. The syntax takes charge of this interface. But linguistic understanding and processing do not end at that point.

On the one hand, the relation between a certain LF (an abstract characterization of the structural meaning of a sentence) and other subsequent levels of representation, either called LF2, level of utterance interpretation or level of mental representation, remains unexplained. This interaction for orthodox modularists (cf., Fodor, Chomsky) is excluded from the language module, whereas according to other cognitivists, linguistic processing specifically includes the building of full semantic representations. In (9) this divergence is synthesized:

> (9) Orthodox modularists are concerned to draw the boundaries of the language input module (and thus all specifically linguistic processing) somewhere below the level of semantic representation, claiming that that module delivers something like a LF representation to central processes, which then interpret it. Marslen-Wilson and Tyler argue that specifically linguistic processing prominently includes the construction of semantic representations (cf., Garfield (1987:22)).

On the other hand, I believe that it is precisely under this architecture of systems of principles (postulated to represent linguistic information) where two more crucial theoretical questions should be posed.

First, we have to decide whether the modularity view is extensive to pragmatic theory. And, second, we have to emphasize the need—which has been developed in the last fifteen years—to incorporate within the system of grammar certain theories which aim at explaining certain non-linear linguistic phenomena that, otherwise, could not be satisfactorily approached.

With regard to the first question, grammar and pragmatics seem to be similar in two respects (cf., Wilson and Sperber (1986:67)): they both fall entirely within the field of cognitive psychology and, moreover, they have to do with language. But, in fact, neither of these aspects implies that they also share modularity.

Some linguists have conceived pragmatics as a specific module; that is, they claim that there is a pragmatic code or system of principles, strategies or rules responsible for the mapping between the output of the grammar and the final utterance interpretation.

The problem with this view is that utterance interpretation cannot be estimated by means of a single code or system of principles, because it is an area where different sets of information (not only linguistic) and principles (not only grammatical) come into play.

According to such a conception, mutual knowledge is necessarily involved and, in fact, it cannot exist, because a model for utterance comprehension based on the decoding of messages in contexts totally fixed before the comprehension process begins is simply mistaken.[1]

In order to make up for these shortcomings, other theorists have thought of pragmatics as a set of hypotheses on the central cognitive systems and on the relation of natural languages to central processes. Therefore, pragmatics is understood to be a set of specific hypotheses on the correlation between grammar and cognition.

In accordance with this line of research, Dan Sperber and Deirdre Wilson argue that pragmatics is just the domain where grammar, logic and memory interact. Pragmatics, according to this view, is not an independent module, but a theory of cognitive principles and a theory on required abilities in utterance interpretation; it is also a theory which focuses on the cognitive effects of utterances.

(10) There is thus a fundamental difference between the modular and the non-modular approaches to pragmatics. A fundamental assumption of the non-modular approach is that the interpretation of communicative behaviour, like the interpretation of evidence in general, is always subject to risk. There are always alternative ways of interpreting a given piece of evidence, even when all the correct procedures for interpretation are applied. These procedures may yield a best hypothesis, but even the best hypothesis may not be the correct, i.e., the intended one. By contrast, decoding procedures, when correctly applied to an undistorted signal, guarantee the recovery not only of *an* interpretation, but of the *correct*, i.e., the intended interpretation. The two approaches start from radically different assumptions about the nature of communication itself (cf., Wilson and Sperber (1986:70-71)).

In the analysis of disjunct constituents which I shall present in the following sections, pragmatics can only be understood as a non-modular system.

As far as the modularity thesis is concerned, another theoretical proposal which constitutes a land-mark in the representation of linguistic information has been the incorporation of non-linear theories into grammar.

Within GG, grammatical theory is conceived as a set of principles that describe and explain the properties of the different levels of representation a given sentence of a language may be associated with and, furthermore, they account for the relationships between all these distinct levels.

In (11) we have the classical triangular schema with the various levels of representation which have been postulated, while in (12) we enumerate some subsystems of principles that control the operation of the linguistic rules (lexical, syntactic, semantic or phonological) and the structure of the representations.

(11) D-Structure
|
S-Structure
/\
PF LF

(12) Binding Theory
Government Theory
Case Theory
Control Theory
θ - Theory
etc.

Moreover, the grammatical model known as Government and Binding Theory has been complemented with various representative or non-linear theories:

(13) Autosegmental Phonology
Metrical Theory
Autolexical Syntax
Multidimensional Syntax [2]

Of particular interest concerning these proposals is the fact that, starting from a given basic assumption (according to which grammar is a modular system with different components, characterized by its autonomy and by the fact that each one of its components is governed by its own principles), they permit the extension of the set of possible relationships among distinct formal representations, since their principles attempt to explain the interaction and integration between representations governed by distinct components of the grammar. The general aim is to develop a grammar of the interface between different types of fully autonomous grammatical structures and, with this new perspective, guarantee their unification.

Thus, a common feature of this sort of theories is the introduction of certain conditions in the way in which linguistic representations are associated with one another.

In sum, it is important to bear in mind that in linguistic theory some theoretical proposals explicitly expressing some sort of linguistic interdisciplinarity (in the strictest sense of the word) already exist: when we face a linguistic phenomenon which, not admitting a linear representation, requires an autonomous structure that in turn has to be synchronized with another linguistic structure.

IV. INTERDISCIPLINARITY IN THE ANALYSIS OF LINGUISTIC INFORMATION

In addition to what I have said so far, certain linguistic phenomena favour the introduction of interdisciplinary approaches, in as much as they pose some problems which affect the organization of the mind and involve the interaction between various units, some of which are linguistic (lexical items, sentences, utterances) while others are not (beliefs, contextual assumptions, etc.).

This means that a new difficulty arises when we wish to know whether certain linguistic data, i.e., sequences of utterances, are going to throw any light on linguistic structures or not. Consequently, one of the problems linguists have to face is deciding whether these data are exclusively the output of the language faculty or whether, on the contrary, they involve distinct faculties and types of knowledge.

On this particular occasion I am going to consider the problem raised by adverbs which are oriented towards the speaker or hearer, in examples such as those in (1b) and (2b). By the end of the paper we ought to be able to answer the following questions:

(14) (a) Which levels of linguistic representation determine which interpretive properties?
(b) Which information is external to the linguistic representation which welcomes the adverb?
(c) Do the utterances with parentheticals provide evidence which might be theoretically relevant for the representation of linguistic information?
(d) Do these utterances pose a syntactic or an interpretive problem?

The need to find an answer to these questions also arises as a consequence of the existence of philosophical contributions (mainly, from the Theory of Speech Acts) that describe and explain the interpretation of disjunct adverbials. Hornstein (1987) observed that:

(15) Much work in linguistic semantics draws considerable inspiration from work in philosophical logic. Concepts and techniques are often transplanted from the latter to the former quite uncritically (cf., Hornstein (1987:146)).

My intention is, therefore, to demonstrate why the incorporation of the Speech Act Theory into the grammatical analysis of utterances containing parenthetical constituents does not prove to be useful, if what we—as linguists—wish to do is to understand a specific aspect of the way in which natural languages are structured: how linguistic information corresponding to parenthetical constituents is to be represented and how disjunct adverbials are to be interpreted.

V. DISJUNCT ADVERBIALS PRAGMATICALLY ORIENTED TOWARDS THE SPEAKER OR HEARER

The most outstanding characteristic of those items oriented towards the speaker or hearer, as in the case of certain adverbs, is that neither are they semantically related to any sentential proposition, nor do they modify any sentential constituent, nor do they predicate over any nominal constituent of the sentence. Moreover, these items have no incidence on the set of truth conditions assigned to the proposition.

Notice that the adverbs presented in (16) are not involved in the propositional content of the sentence, since the corresponding semantic entities are not relevant at all to the set of conditions which describe the proposition.

(16) (a) Personalment, considero que has fracassat
(= Personally, I believe that you've failed)
(b) Considero, sincerament, que has fracassat
(= I sincerely believe that you've failed)
(c) Considero que, francament, has fracassat
(= Frankly, I believe that you've failed)
(d) Considero que has fracassat, honestament
(= I honestly believe that you've failed)

Furthermore, the reader should pay attention to the fact that these items, although they are neither predicates nor operators, do contribute to the information transmitted by the utterance: they are understood as linguistic markers which inform about the speaker / hearer involvement in the communication act.

On the other hand, it is unquestionable that a range of constituents with different syntactic configurations may express the speaker's point of view or attitude towards the utterance. So, the pragmatic orientation towards the speaker / hearer is not an exclusive property of certain adverbials. Some examples which illustrate this aspect are given in (17):

(17) (a) En la meva opinió, considero que has fracassat
(= In my opinion, I believe that you've failed)
(b) Si vols que et digui la veritat, ...
(= To tell you the truth, ...)
(c) Per a ser honest, ...
(= To be honest, ...)
(d) Amb la mà al cor, ...
(= In all honesty, ...)

Going back to our examples in (16), some grammarians who have been working within a generative framework (cf., Schreiber (1972) and Lonzi (1981)) have postulated certain types of abstract representations (with a performative verb in present indicative tense, a first person subject, and a second person indirect object), where the adverb is interpreted as a modifier of a speech act that is explicitly expressed at an abstract level of syntactic representation.

The Performative Hypothesis Incorporated into Grammatical Theory

The attempt to provide a syntactic basis for the interpretation of disjunct adverbials goes hand in hand with the hypothesis that all sentences (declarative, interrogative and imperative) have performative structures in their underlying representations and, therefore, that any sentence in (16) also has an underlying structure with performative markers. (18) corresponds to (16c).

(18) [Et dic] francament [que] considero que has fracassat

Schreiber (1972) explored the performative hypothesis by means of its incorporation into grammar, and he postulated a sort of fully abstract D-structures which would include constituents not recoverable from S-structure. As a result, he also assumed a Rule of Performative Deletion that in (18) would delete all the items preceding the adverb and the conjunction *que*. (16c) would be the output structure of that rule and a rule of adverb movement.

Likewise, under this syntactic approach to the role of the speaker/hearer in communication acts, those sentences in (19) would be considered as semantic paraphrases and they would correspond to a similar D-structure (something like (19c)).

(19) (a) Honestamente, estás equivocado
(= Honestly, you are wrong)
(b) Te digo honestamente que estás equivocado
(= I tell you honestly that you are wrong)
(c) Soy honesto al decirte que estás equivocado

(= I'm honest in telling you that you are wrong)

Notice that Schreiber's proposal entails a complex derivational process: the adverb would stem from a functionally distinct adverbial form, which in turn would probably derive from an adjectival form working as a main predicate.

(20) adjective --> manner adverb --> disjunct adverbial

Naturally, Schreiber's starting point is the supposed synonymy between the three propositions in (19); but this paraphrase relation, which is already questionable with respect to (19b-c), is still more dubious between these two propositions and (19a), whose propositional content is undoubtedly less informative, if we just consider the strictly linguistic information it contains.

Similarly, Lonzi (1981) syntactically relates an utterance such as the one in (21a) to the abstract structure in (21b):

(21) (a) I promise you, confidentially, that the manager will resign
(= Os prometo, confidencialmente, que el director dimitirá)
(b) [[[I promise you that the manager will resign]
E S' S'
[PRO speaking confidentially]]]
S'

According to her proposal a sentential structure is subordinated to E (= expression, utterance), which represents the linguistic act. In (21b) PRO is the subject of the gerund, obligatorily controlled by the subject of the performative verb. *Speaking* is an abstract verbal form which is modified by a derived adverb.

The main problem of both studies is that the arguments adduced are more intuitive than convincing, since they rest on a supposed and inexact paraphrase relation between sentences with performative verbs and sentences with adverbs.

On the other hand, taking into consideration some ideas already discussed in the introduction, it is not at all obvious that everything manifested in language is due to the operation of the faculty of language and, therefore, that everything manifested through language has to described by the theory of grammar.

Besides, these analyses imply an increase in the generative power of the transformational rules required in grammar. And, moreover, they leave other utterances, which also contain disjunct adverbials (or parenthetical constituents of a different syntactic nature), but which under no circumstances can be interpreted as conditions on a certain utterance or speech act, unexplained. In other words, without any explicit marker, certain linguists go too far in trying to force a propositional reading. This is what the examples in (1b) and (7) suggest.

In the remainder of this section I shall try to demonstrate the shortcomings of these works and I shall argue against the incorporation of the philosophical Theory of Speech Acts into the syntactic representation of the utterances under study.

A. According to those linguists that believe that the speech act theory can solve the problem of meaning posed by some linguistic utterances, the central task of semantics consists in explaining the meaning of the sentence, not in terms of its internal relations (what would correspond to the principles of the LF- component), but in terms of what a speaker is able to do with a sentence: which act he fulfils and which knowledge (or presuppositions) he assumes in the fulfilment of the corresponding act.

Assuming the performative hypothesis, the role of the speaker / hearer is retrieved from the abstract performative structure which is postulated and, more accurately, by specifying the subject of the clause whose abstract verb is modified by the manner adverb before the rule of performative deletion is applied. Thus, for example, the interpretation of the two sequences in (22) would be made explicit from those underlying structures in (23).

(22) (a) Sincerament, em vas mentir
(lit.= Sincerely, you lied to me)
(b) Sincerament, em vas mentir?
(= Sincerely, did you lie to me?)

(23) (a) Et dic *sincerament* que *em vas mentir*
(= I tell you sincerely that you lied to me)
(b) Et demano que em diguis *sincerament* si *em vas mentir*
(= I ask you to tell me sincerely whether you lied to me)

Notice that structures such as those underlying (23) do not delimit the meaning or the information the adverb transmits; rather, they are an approximation to the presuppositions of the speaker at the very moment of utterance production, an approximation to those assumptions that the hearer ought to know (or infer) in order to accurately understand the corresponding utterance. Whatever the case, these presuppositions do not provide any information on the linguistic function of the adverbial constituent with respect to the proposition it goes with, under the assumption that some specific relation exists.

The problem with such an approach is that it only aims at the introduction—within the sentential syntactic representation—of the adverbial pragmatic orientation. And notice that, this being the situation, semantic theory is on a level with performance theory and should be banished from the language module.

In fact, there is no conflict between a semantic theory subject to specific well-formedness conditions on anaphoric relations, thematic relations and scope relations, and a semantic theory on utterance understanding based on the notion of speech act, and integrated within a performance theory. Moreover, a speech act theory is not only left outside the language module, but its theoretical apparatus is inadequate to describe the information transmitted by many other disjunct constituents (see the examples in (1) to (3) and (7)).

B. A second crucial problem is that an interpretive analysis of adverbials oriented towards the speaker / hearer in terms of the concept of speech act has forced the introduction of underlying representations as abstract as of dubious accuracy (see, for example, the syntactic representation in (21b)).

An interpretive problem has changed into a problem of syntactic representation, and this approach has entailed a number of theoretical drawbacks, among which the need for deletion rules has to be pointed out. Remember that in the present state of the theory this kind of rules—since they do not leave any trace that permits the irrecoverability—have been eliminated from linguistic analyses. Hence, in spite of the fact that in the syntactic derivation of performative sentences a deletion of certain syntactic material has been assumed, these powerful operations cannot exist because they do not obey the set of general principles (cf., Government Theory and Trace Theory) which control the Move—α schema.

C. Finally, I would like to point out that the analysis developed by Schreiber (1972) and Lonzi (1981), which consists in incorporating a pragmatically oriented hypothesis within the linguistic description of disjunct adverbials, has been based on a series of false arguments.

C.1. Schreiber claims that the ungrammaticality or malformation resulting from the adjunction of disjunct adverbials into subordinate structures (look at (24a)) is accounted for if, and only if, disjuncts are derived from manner adverbials that would have modified a performative clause already deleted. His argument is based on the assumption that performative clauses can never appear freely within subordinate structures.

Certainly, the examples in (24) are badly formed:

(24) (a) # Los científicos me informaron de que, personalmen te, el dinosaurio se ha extinguido
(lit. = The scientists informed me that, personally, the dinosaur is extinct)
(b) # Los científicos me informaron de que te digo personalmente que el dinosaurio se ha extinguido
(= The scientists informed me that I tell you personally that the dinosaur is extinct)

However, notice that (25) is as odd as (24) is:

(25) # Personalmente, el dinosaurio se ha extinguido
(= Personally, the dinosaur is extinct)

and here the adverb *personalment* is not in a subordinate position, which means that the cause of its ungrammaticality, if one can speak of ungrammaticality, should be one other than that pointed to by Schreiber.

In addition, another set of examples shows that disjuncts may arise within a subordinate clause. Notice that one may not talk of malformation in any of the following utterances:

(26) (a) Les he comunicado que, seriamente, no me interesan los cargos
(lit. = I've told them that, seriously, I'm not interested in any post of responsibility)
(b) Lola Flores ha declarado que, personalmente, no era consciente de que hacía un fraude
(lit. = L.F. has declared that, personally, she was not aware she was committing fraud)
(c) Ahora resulta que me aseguras que, sinceramente, la falta de ozono puede ser muy perjudicial
(lit. = Now the fact is that you assure me that, sincerely, the lack of ozone may be very harmful)[3]

In my view the malformation of (24) and (25) is not due to a problem of syntax but to an informative antagonism between the content of the disjunct adverbial *personalment* and the propositional content corresponding to *el dinosaurio se ha extinguido*; that is, it does not seem relevant to present the informative content of a scientific truth (synthetic, contingent and a posteiori) as a personal opinion.

C.2. Schreiber also claims that the ungrammaticality which results from the juxtaposition in a fixed order of two adverbial lexemes (one modal and the other disjunct) ought to be paired with the restrictions on the basic order of adverbials within a performative representation.

Hence, Schreiber relates the malformation of (27) with the fact that, following a performative analysis, the underlying order should be the one in (28).

(27) (a) # Evidentemente, honestamente, la ayuda económica ha sido muy discreta
(= Evidently, honestly, the economic aid has been very limited)
(b) # Posiblemente, francamente, los miembros de la comisión siempre coinciden
(= Possibly, frankly, the members of the comission always agree)

(28) (a) Te digo honestamente que es evidente que la ayuda económica ha sido muy discreta
(= I tell you honestly that it is evident that ...)

(b) Te digo francamente que es posible que los miembros de la comisión siempre coincidan
(= I tell you frankly that it is possible that ...)

Notwithstanding, Schreiber does not seem to have realized that the sentences in (29), which show exactly the same superficial adverbial order as in (27), should also refer to the underlying order given in (28),but—in spite of that—they are fully grammatical.

(29) (a) Evidentemente, la ayuda económica ha sido, honestamente, muy discreta
(= Evidently, the economic aid has, honestly, been very limited)
(b) Posiblemente, los miembros de la comisión, francamente, siempre coinciden
(= Possibly, the members of the comission, frankly, always agree)

The explanation of (27)'s malformation cannot be, once again, what Schreiber (1972) suggested.

Furthermore, if we consider the utterances in (30):

(30) (a) ? Honestamente, evidentemente, la ayuda económica ha sido muy discreta
(= Honestly, evidently, the economic aid has been very limited)
(b) ? Francamente, posiblemente, los miembros de la comisión siempre coinciden
(= Frankly, possibly, the members of the comission always agree)

we notice that they are much better than those in (27) and, presumably, this is due to conditions on language processing. So, I would not adduce anything like ungrammaticality in (27) and grammaticality in (30), since all examples seem to obey the principles of grammar.[4]

Finally, notice that amongst (30) and (29) any Spanish native speaker/hearer would prefer those linguistic sequences where two *-ly* adverbs are not adjacent in linear order.

C.3. A third argument posed by Schreiber, to derive disjunct adverbials from underlying structures with performative verbs, arises from sequences where one attempts to negate a disjunct adverbial.

(31) (a) # No seriamente, ya se ha terminado el curso
(= Not seriously, the term has already finished)
(b) # No honradamente, pensaba no presentarme a las elecciones
(= Not honestly, I was not thinking of running at the elections)

His argumentation is as follows: a performative analysis of disjunct adverbials obviates the need to treat the malformation of (31) by turning to an "ad hoc" restriction, since the underlying structures they would derive from are also badly formed.

(32) (a) # Te digo no seriamente que ya se ha terminado el curso
(= I tell you not seriously that ...)
(b) # Os comunico no honradamente que pensaba no presentarme a las elecciones
(= I tell you not honestly that ...)

Again, I am inclined to suggest that we do not face a problem of syntax, for adverbial negation is feasible in sentences such as those in (33).

(33) (a) NO SERIAMENTE, criticaban los diputados la gestión del ministro

(lit. = NOT SERIOUSLY, were the deputies criticizing the minister's work)

(b) NO HONRADAMENTE, había examinado el abogado el legado del pintor
(lit. = NOT HONESTLY, had the lawyer examined the painter's legacy)

These data, which illustrate the possibility of emphasizing exactly the same negated adverbial constituents we find in (31), seem to suggest that the sequences in (31) are not syntactically ungrammatical. The syntactic component of grammar ought to generate these sequences and, therefore, the fact that they are anomalous can hardly be attributed to any restriction within the language module. I think, rather, that—when dealing with those sequences in (31)—we should think of a violation of certain principles of communication: when the negative adverb operates on adverbials which are used to manifest the beliefs or attitude of the speaker towards the truth of *p*, then the utterances obtained violate the maxim of quality proposed by Grice (1975) and, obviously, they also violate the principle of maximal relevance argued for by Sperber and Wilson (1986).

An utterance is said to be relevant if the hearer is able to enlarge or modify his initial set of beliefs or assumptions. Now then, the premisses which are incorporated into the hearer's contextual knowledge can only be those propositions which are true or the hearer believes, assumes or has some evidence to suppose that they are true, and in (31) the negation of the adverb obstructs the relevance of the utterance.

To sum up, I have so far been dealing with different problems that emerge as a result of the incorporation of the philosophical theory of speech acts into the syntactic analysis of linguistic sequences containing disjunct adverbials.

The examples given in (1) to (3) and (7) provide some evidence that not all disjunct constituents, not even all disjunct adverbials, are contextualized in such a way that the assumption of a modified constituent, identified with some proposition recovered from context,is inferred. We could certainly postulate that disjunct adverbials such as those in (1b) and (2b) do introduce a metalinguistic condition on a specific speech act only recoverable from context. But, notice that in examples similar to (7) the temporal AdvP does not modify the propositional content of the sentence *recordo la petita A.* it appears with, nor any other proposition retrievable at the level of utterance understanding. The adverbial construction seems rather to guide the hearer towards a conceptual representation which is itself saturated: the image of little A. last autumn sucking, with her eyes closed, a wet golden grape. This means that some adverbs may end up without modifying, in the sense we understand this term in linguistic theory, any proposition recovered from linguistic or non-linguistic context.

Another objection refers to nominal disjuncts. What should be said about the syntactic representation and LF of nominal disjuncts such as those in (1b) and (7)? Should we assume that, since they are nominal, they should correspond to nominal arguments? And, do they have to be assigned a thematic role at the postsyntactic level of utterance interpretation? I have the feeling that this sort of reasoning would lead us too far away from what may be considered to be an appropriate model for the representation of linguistic information.

I would say that it is up to the hearer to decide whether *noia* in (1b) and *catorze anys* in (7) have to correspond to fully saturated constituents within his/her set of thoughts and assumptions, or whether they have to be paired with a conceptual structure with propositional form. The key to the problem is that disjunct constituents may correspond to mental representations that are not linguistically saturated as predicates, arguments or operators.

In conclusion, we must provide another type of answer (distinct from the performative hypothesis) to the question of what is the syntactic representation and LF corresponding to utterances containing parenthetical information, either meta communicative information over some active participant in an act of verbal communication or some other type of information.

VI. ON THE REPRESENTATION OF DISJUNCT CONSTITUENTS:A MULTIDIMENTIONAL APPROACH

Up till now we have been able to see that the main problem posed by utterances with disjunct constituents—a problem that is still unsolved—is the question of how parenthetical constituents should be represented with respect to those sentences they interact with: how non-restrictive relative clauses, clauses with an independent intonational structure, parenthetical phrases or—more precisely, in linguistic terms—disjunct constituents should be treated in linguistic theory. The examples introduced so far have illustrated the fact that disjunct constituents may belong to different syntactic categories. The property of being a disjunct is not exclusive of adverbials, nor of adverbial phrases or adverbial clauses, since it is also a potential property of NPs, relative clauses, certain independent sentences and other major syntactic categories.

Some works, as yet unpublished (cf., Espinal (1985, forthcoming) and Haegeman (1988)), bring forward a series of arguments that substantiate the thesis that disjuncts are independent constituents, in the widest sense of the term.

Syntactically, they do not participate in syntactic processes taking place within the syntactic domain of the sentence: they cannot be the focus of a cleft sentence; they cannot be questioned; they are not under the effects of certain restrictions in the combination of verbal tenses; they cannot be substituted for any valence requirement and more than one disjunct constituent may be associated with a sentential structure.

Semantically, they do not have any thematic relation with any argument of the syntactic structure in which they appear, they do not modify the predicate or the whole propositional sentence, nor are they modified by a propositional operator (the negation or other temporal operators). At the linguistic level of LF, they are neither logical predicates nor logical operators. This would imply that the meaning of parenthetical expressions is not integrated within the formal representation of the meaning of the sentence.

Phonetically, they are always associated with an intonation unit independent of the one associated with the sentence in which they occur.

Finally, all the examples show that, pragmatically, disjunct constituents either connect with the speaker or addressee, or they modify an implicit speech act, or they introduce an additional assumption of the speaker into the discourse. Disjunct constituents are units of information disassociated from the proposition with which they have to be simultaneously interpreted at the moment of utterance processing.

In addition to those properties, it has been observed that the linear order of disjunct constituents has no effect—in the unmarked cases—on the grammaticality of the syntactic structure in which it occurs, nor on its meaning.

This set of considerations has led me to propose the hypothesis, which I shall briefly outline in the remaining part of this paper, that disjunct constituents have their own syntactic structure. Therefore, in order to answer the question of how to represent the information transmitted by a linguistic expression which is not formally integrated within the sentential structure where it occurs, I wish to postulate that linguistic utterances may distribute their syntactic information across various syntactic isochronic representations.

(34) The representation of disjunct constituents coocurring with sentences consists of n-dimensional syntactic structures

I believe that we have crucial evidence to implement the grammatical model in such a way that a grammatical linguistic sequence might correspond to a multidimensional syntactic representation. I shall assume, accordingly, that in a non-linear theory, the syntactic description of complex utterances with disjunct constituents will be translated into a number of independent syntactic structures linked by means of different mechanisms.

(35) Different and independent syntactic representations may conflate in a given utterance at a certain time. Linguistic utterances may be represented in one single structure or in n-dimensional isochronic formal representations. In the first case, the syntactic structure is standardly conceived as being of a sentential type. In the second case, $n > 1$ syntactic structures can be sentential or not.

A set of grammatical and pragmatic principles, as well as some well-formedness conditions, will control the association between the different phrase markers.[5]

Under this new approach to the representation of linguistic information, (36a) would correspond to (36b), where the three independent phrase markers should be understood as representing different structural dimensions.

(36) (a) El conferenciante llegará -me imagino que ¡ojalá sea así! en avión- el próximo martes
(= The speaker will come -I guess, let's hope so!, by plane- next Tuesday)

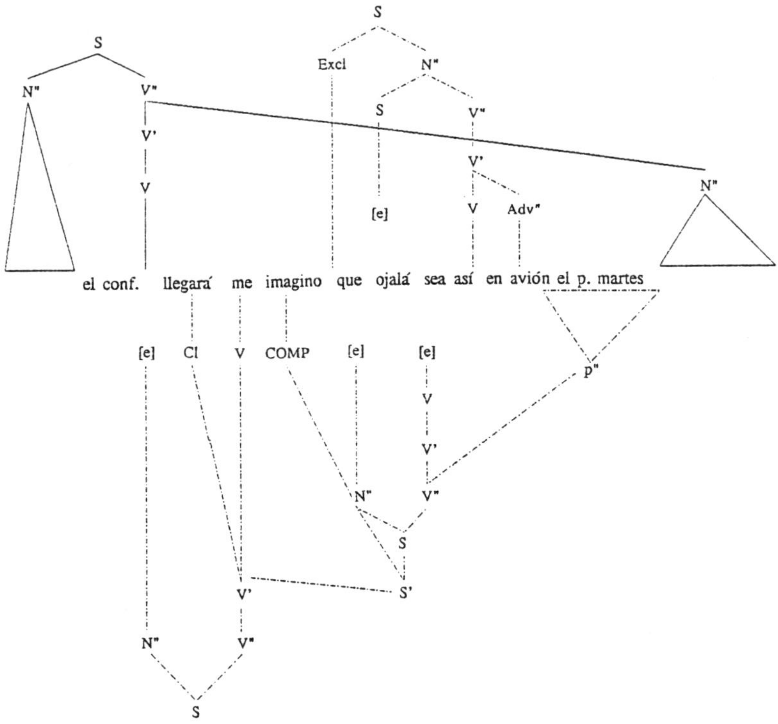

Notice that a non-linear syntactic model for disjuncts will predict some features of the syntax and semantics of disjunct constituents.

First, the postulated structural independence between different phrase markers accounts for the linguistic independence between disjuncts and host sentences, and also for the non- configurational property of those sequences with disjunct adverbials.[6]

Second, a multidimensional syntactic approach is the first step towards an explanation of the independence of meaning between the various phrase markers. Each phrase marker corresponds to an independent LF representation and every construct of each one of them must be externally licensed by being assigned an interpretation appropriate to its syntactic category. What is empirically incorrect is the assumption that all the information contained in a complex linguistic sequence with a disjunct constituent and a fully saturated sentence is going to be satisfied at a single LF for the whole sequence. At this level of the interface between grammar and cognition we know as LF, the semantic information of disjuncts is on some occasions going to depend on its own syntactic structure and on others is going to be reduced to the semantic information specified in the lexicon.

Furthermore, in order to explain how in a linguistic processing act the unification or incorporation of information between the various syntactic markers takes place, the work of some syntactic and conceptual well-formedness conditions is needed. These syntactic and conceptual association rules will guarantee that disjunct constituents end up at the right place and will control the mapping between major syntactic constituents and conceptual constituents (cf., Jackendoff (1983, 1987) and Lakoff (1986a, 1986b)).

Finally, in this utterance processing operation, which is meant to find a way of saturating the different conceptual categories corresponding to a verbal message, one may require not only what is codified through the linguistic input, but also what is percieved from the external environment and what is assumed in one's encyclopaedic knowledge. This is the point at which pragmatic association principles are involved: the concept of relevance imposes a constraint on the balance between the processing cost and inferential reward of an utterance containing disjunct constituents (cf., Sperber and Wilson (1986), Kempson (1988b)).

VII. CONCLUSIONS

(37) (a) Linguistic utterances containing disjunct constituents seem to correspond to multidimensional syntactic representations

(b) The assignment of an autonomous syntactic representation to a disjunct constituent and its interpretation depend on the accessibility of contextual knowledge as well as on the accessibility of universal conceptual and pragmatic principles, which are outside the language module and control the interface between grammar and cognition

(c) The conceptual information of disjunct constituents is neither encapsulated within a single sentential structure nor isolated from the central cognitive system. There is no compositional interpretation for the disjuncts

The syntactic information corresponding to a grammatical linguistic expression does not necessarily have to meet in a single phrase marker; rather it may be multidimensional. This means that the various dimensions (the phrase marker corresponding to the host sentence and the phrase marker(s) of the disjunct constituent(s)) codify autonomous syntactic information. This is how linguistic utterances containing disjuncts would be grammatically represented.

In addition, to account for the information transmitted by an utterance containing a disjunct syntactic constituent, neither this syntactic analysis nor the set of principles of the LF- component are sufficient, because the interpretation of a disjunct constituent requires the hearer to have access to some contextual knowledge, as well as to some conceptual and pragmatic universal principles that are outside the language module and that control the grammar - cognition interface.

As has been shown, some linguistic utterances exist whose information is neither encapsulated within a unique sentential structure at any level of linguistic representation nor isolated from the central cognitive system. In other words, the linguistic meaning is not always compositional and, in order to be able to identify the level of representation at which the meaning of the disjunct is unified with the propositional meaning, grammatical principles do not suffice.

Finally, notice that a crucial advantage of this approach is that it does not increase the generative power of grammar. All the proposals one may find in the linguistic literature that discuss the representation of disjunct information add special mechanisms to grammatical theory. Schreiber (1972) and Lonzi (1981), as we have already seen, encapsulate the parenthetical information of disjunct adverbials pragmatically oriented towards the speaker / hearer within an abstract performative structure which must subsequently be submitted to a deletion process. Other proposals increase either the degrees of hierarchic embedding within a syntactic structure (cf., Jackendoff (1977)), or the number of linguistic levels of representation (cf., Safir (1986)), or the range of necessary empty categories (cf., Haegeman (1988)).

On the contrary, the analysis here outlined relies on the assumption that a set of independently motivated syntactic, conceptual and pragmatic association rules are going to unify, although not encapsulate, the set of phrase markers.

NOTES

*. This paper has benefited from the advice and encouragement of Víctor Sánchez de Zavala. Thanks also to Allison Beeby and Olivia Fox for comments on an earlier English version.

1. It is mistaken because it assumes that in language understanding any task is easily programmed and predictable: reference assignment, disambiguation, attitude retrieval, figurative and other implicit interpretations do not work mechanically, through the application of a set of rules which automatically integrate a fixed context with the semantic properties of the utterance (cf., Sperber and Wilson (1982)).

2. Autosegmental Phonology or Multilinear Phonological Theory (cf.,Goldsmith (1976)) was designed to deal with linguistic phenomena such as tone, intonation, vowel and consonant harmony, gemination and compensatory lengthening. Metrical Phonology or Hierarchic Phonological Theory (cf., Liberman and Prince (1977)) was postulated to account for the problem of stress. Autolexical Syntactic Theory (cf., Sadock (1985)) was introduced to explain some cliticization phenomena and nominal incorporation. And, Multidimensional Syntactic Theory (cf., Goodall (1984), Haegeman and Van Riemsdijk (1986)) was originally proposed to solve the analysis of coordination structures and verbal reanalysis.

3. Note the fact that (26b) proves the common assumption that disjunct adverbials are pragmatically oriented either towards the speaker or towards the hearer is false: in fact, they can also be oriented towards the third person.

4. Apparently, the problem posed by (27) seems to be that the expectations introduced by the modal adverb (which works as a propositional operator at the level of LF) are seriously cut off by a pragmatically oriented disjunct before the sentential

proposition begins. However, the following examples, which contain other disjuncts, are completely well formed:

(1) (a) Evidentemente, yo diría, la ayuda económica ha sido muy discreta
(= Evidently, I would say, the economic aid has been very limited)
(b) Posiblemente, ustedes recordarán, los miembros de la comisión siempre coinciden

(= Possibly, you will remember, the members of the comission always agree)

5. In another paper (cf., Espinal (forthcoming)) the arguments for a multidimensional syntactic approach to disjuncts and the work of the syntactic, conceptual and pragmatic association rules are discussed in detail.

6. In spite of the fact that utterances show superficial linearity, this does not entail any kind of linear structure between the disjunct constituent and any constituent of the host sentence in the syntax.

REFERENCES

Carston, R. (1988), "Language and cognition," in F. Newmeyer (ed.), *Linguistics: the Cambridge survey III. Language: psychological and biological aspects*, Cambridge University Press, Cambridge.

Espinal, M. T. (1985), *Anàlisis interpretatives i teoria lingüística*, Tesis Doctoral, Universidad Autónoma de Barcelona, Bellaterra.

Espinal, M. T. (forthcoming), "On the representation of disjunct constituents: an approach to multidimensional syntactic structures," Ms. Universidad Autónoma de Barcelona, Bellaterra.

Fodor, J. A. (1983), *The modularity of mind*, The MIT Press, Cambridge Mass.

Garfield, J. L. (ed.) (1987), *Modularity in knowledge representation and natural-language understanding*, The MIT Press, Cambridge Mass.

Goldsmith, J. (1976), "An overview of autosegmental phonology," *Linguistic Analysis*, 2, 23–68.

Goodall, G. (1984), *Parallel structures in syntax*, Doctoral dissertation, University of California at San Diego; Cambridge University Press, Cambridge, 1987.

Grice, H. P. (1975), "Logic and conversation," en P. Cole and J. L. Morgan (eds.), *Syntax and semantics 3: Speech acts*, Academic Press, New York.

Haegeman, L. (1988), "Parenthetical adverbials: the radical orphanage approach," LAGB spring meeting, Durham.

Haegeman, L. and H. van Riemsdijk (1986), "Verb projection raising, scope and the ty–ology of verb movement rules," *Linguistic Inquiry*, 17, 417–466.

Hornstein, N. (1986), "Pragmatics and grammatical theory," *Chicago Linguistics Society*, 22, 234–247.

Hornstein, N. (1987), "Levels of meaning," in J. L. Garfield (ed.).

Jackendoff, R. S. (1977), *X' syntax: a study of phrase structure*, The MIT Press, Cambridge Mass.

Jackendoff, R. S. (1983), *Semantics and cognition*, The MIT Press, Cambridge Mass.

Jackendoff, R. S. (1987), "X-bar semantics," *Berkeley Linguistics Society*, 13, 355–365.

Kempson, R. (1988a), "Logical form: the grammar cognition interface," *Journal of Linguistics*, 24, 393–431.

Kempson, R. (1988b) "On the grammar - cognition interface: the principle of full interpretation," in R. Kempson (ed.), *Mental representations. The interface between language and reality*, Cambridge University Press, Cambridge.

Lakoff, G. (1986a), *Women, fire and dangerous things*, University of Chicago Press, Chicago.

Lakoff, G. (1986b), "Frame semantics, control and the coordinate structure constraint," *Chicago Linguistics Society*, 22, 152–167.

Liberman, M. and A. Prince (1977), "On stress and linguistic rhythm," *Linguistic Inquiry*, 8, 249–336.

Lonzi, L. (1981), "Avverbi frasali e strutture parentetiche," *Lingua e Stile*, 16.3, 393–431.

Marslen-Wilson, W. and L. K. Tyler (1987), "Against modularity," in J. L. Garfield (ed.).

Pollard, C. and I. Sag (1987), *Information-based syntax and semantics*, CSLI, Stanford.

Sadock, J. M. (1985), "Autolexical syntax: a proposal for the treatment of noun incorporation and similar phenomena," *Natural Language and Linguistic Theory*, 3, 379–439.

Safir, K. (1986), "Relative clauses in a theory of binding and levels," *Linguistic Inquiry*, 17.4, 663–689.

Schreiber, P. A. (1972), "Style disjuncts and the performative analysis," *Linguistic Inquiry*, 3, 321–347.

Sperber, D. and D. Wilson (1982), "Mutual knowledge and relevance in theories of comprehension," in N. Smith (ed.), *Mutual knowledge*, Academic Press, London.

Sperber, D. and D. Wilson (1986), *Relevance communication and cognition*, Blackwell, Oxford.

Urmson, J. O. (1952), "Parenthetical verbs," *Mind*, LXI, 480–496.

Wilson, D. and D. Sperber (1986), "Pragmatics and modularity," *Chicago Linguistics Society*, 22, 67–84.

CHAPTER 5

MODELLING MEMORY FOR MODELS

Keith Stenning
Human Communication Research Centre
Edinburgh University

ABSTRACT

A "model" of a text is a mapping of its constants, predicates and relations onto a set of individuals. Stenning (1975, 1978) observed that powerful conventions for the interpretation of expository text ensure that texts determine models.

Two contrasting theoretical interpretations of these and other observations are current in the literature. Johnson-Laird and his associates (eg., Johnson-Laird & Steedman 1978, Johnson-Laird 1983) interpret models as a *medium* of mental representation. Alternatively, models may be treated as abstract objects which are important because they specify *what is represented.* For this latter interpretation, the question how representation is achieved is the outstanding psychological problem for theories of text comprehension and memory.

A central problem in the representation of models is the representation of individuals and their properties. This paper gives a brief description of the results of a research program aimed directly at the question of how models are represented during and after text comprehension. Analysis of memory performance (Stenning, Shepherd & Levy 1988) reveals that the underlying representation of individuals is distributed through a set of features representing existential facts about the whole ensemble of individuals described. Stenning & Levy (1988) show that a PDP simulation of the representations revealed in these analyses explains human error patterns.

Finally, Johnson-Laird's theory of syllogistic reasoning is reconsidered in the light this interpretation of models. We show that mental models notation is less distinct from other proposals than at first appears, and offers only another logical account. A psychological account of human syllogistic reasoning requires a theory of how the binding of properties to individuals is achieved during reasoning. We show how the theory of distributed representation of binding outlined here can explain why human beings adopt the theorem proving strategies described in mental models.

I. INTRODUCTION

An important shift of emphasis has occurred within cognitive science in recent years away from sole concentration on disembodied abstract systems toward consideration of implementation and resource limitations of implementations. This is good news for psychologists because their main source of data and their main theoretical concerns are with what beings finite in both space and time can know and do. This shift will also improve understanding between formal disciplines such as logic or linguistics and experimental disciplines such as psychology because issues of implementation interpose a buffer between formal system and behaviour, which allows richer explanation of apparent divergences. Cases in which psychologists have claimed that human behaviour flouts the canons of a formal system can be reassessed as breakdowns due to resource limitation. Many psychologists' arguments about the inappropriateness of logic to describing human reasoning are arguments against some tacit implementation of logic. It has been all to easy to forget that logics do not *do* anything until they are implemented as reasoning systems.

The development of logics and grammars was intimately associated with the development of a particular type of computational architecture: the Turing/Von Neumann machine. Architectures make some things easy and others difficult. Different architectures make different things easy and difficult. The type of computational architecture in which we couch our theories has profound effects on which phenomena we choose to analyse. Recent widespread interest in connectionism has had the useful effect of bringing attention to an architecture which has almost the opposite characteristics to the conventional Von Neumann machine. In connectionist systems, multiple parallel soft constraint satisfaction is made easy whereas variable binding is made hard.

For a psychologist interested in the resource limitations exhibited by human reasoning and memory this is important news. These strengths and weaknesses accord better with human strengths and weaknesses than do those of a Von Neumann machine. My purpose here is to show that this new perspective brings to the centre of attention the binding problem in memory. This problem has been thought too low-level to be worthy of cognitive attention, but a solution to this problem can explain features of human reasoning which have been claimed to show that human reasoning has no relation to logic. If these explanations are correct, human syllogistic reasoning is just a straightforward implementation of a classical logic in a system with a very different working memory architecture than that of a von Neumann computer.

The best introduction to the binding problem in working memory is through the anaphor resolution problem, and this introduction illustrates the influence of architecture on the identification of research questions. These two problems are opposite sides of the same coin, but choice of modelling architecture determines which is seen to be a problem.

The anaphor resolution problem is too well known among cognitive scientists to need much introduction. Potential antecedents occur in texts and at later points anaphors follow. Typical anaphors are pronouns and their antecedents may be a full noun phrase or sentences. This is a task which is easy most of the time for human beings, and as with most of our skills, passes unnoticed. It initially became an issue for cognitive psychologists when attempts to get conventional computers to process natural language discourse revealed how difficult this problem is for conventional computational systems. It is a hard problem for such systems because it is knowledge-rich: the correct resolution of an anaphor can potentially depend on almost any arbitrarily inaccessible piece of general knowledge.

The second problem, the binding problem, is in contrast a problem which has received little explicit attention from cognitive scientists until quite recently. It is the problem of how we remember what assignments we made when resolving an anaphor. When we decide that an anaphor "he" actually refers to Tom rather than to Dick or Harry, we have to remember that the property which was predicated of it belongs to Tom rather than to Dick or to Harry. When there are several properties of several people and these properties could, on general knowledge grounds, equally well belong in any combination to any of these individuals, this presents a difficult problem for human memory. Everyone immediately recognises this as something difficult to achieve.

It is a very easy problem of course for conventional computers. Anyone who has had an introduction to a programming language will know a number of ways of solving the binding problem by structural means. It is for this reason that it has received relatively little attention from cognitive psychologists. It is widely seen as cognitively low-level, and if it were solved by human beings the way that conventional computers solve it, it would indeed be low-level. But we have known for at least thirty, and arguably a hundred years, that human beings solve it by knowledge-rich means. I will argue here that a good theory of the content dependence of human working memory's solution to the binding problem would offer an explanation of at least one sort of content dependence of human reasoning. This latter is a problem which has exercised cognitive psychologists a good deal.

For example, Wason & Johnson-Laird (1972) have shown how content dependent human reasoning is, and have argued that this dependence raises challenges for the relation between human reasoning and logic—which is, after all, formulated to give a content independent theory of correct deductive reasoning. Johnson-Laird (1983) has put forward a detailed theory of syllogistic reasoning which purports to be an *ana*logical theory of reasoning and to explain how human reasoning is semantically driven and content dependent. The representations employed in this theory are claimed to displace logic and simultaneously to avoid problems posed by other better-known graphical methods such an Euler's circles and Venn diagrams.

My purpose here is to shift the perceived locus of the problem. The current emphases, I claim, arise out of a misunderstanding about what logic does and does not set out to do. Logic is not an account of a computational process or even architecture which itself predicts what people *do*. In order to fulfill that role, logic requires implementing, and the important issues are about the nature of the implementation which people have. The arguments against the relevance of logic to human mental processes, at least the arguments that concern us here, are arguments against specific types of implementation, namely the ones which conventional computers embody. The difficult psychological task is to reveal the nature of the implementation of logic in people, and here what is primarily, but not solely, required is an account of human working memory architecture.

My plan is to begin by sketching the sort of theory of memory required and by describing, in outline, our work on such a theory to date. I will then briefly consider some of the alternative types of representation which have been claimed to underlie human syllogistic reasoning, and show that they are not as different as has been supposed. Finally, I will show how the theory of memory sketched can potentially explain how people implement classical logic, and why they have the limitations observed. If this account is right, it is an example which shows that the major task facing cognitive psychology is the explanation of human implementations, a task for which choice of architecture for modelling is all important.

II. TWO SENSES OF "MODEL"

The concept of a model is critical for understanding both anaphora resolution and the binding problem. Texts usually determine unique, possibly partial, mappings from properties and relations onto sets of individuals. In fact, the main difference between what are classed as text comprehension problems and what are classed as verbal reasoning problems is just that the latter consist of texts which do not straightforwardly determine the construction of unique models sentence by sentence.

Statement set:

There are three things. Two are white. One is black. Two are hexagonal. One is triangular.

Model description:

There is a triangular white thing, a hexagonal black thing, and a hexagonal white thing.

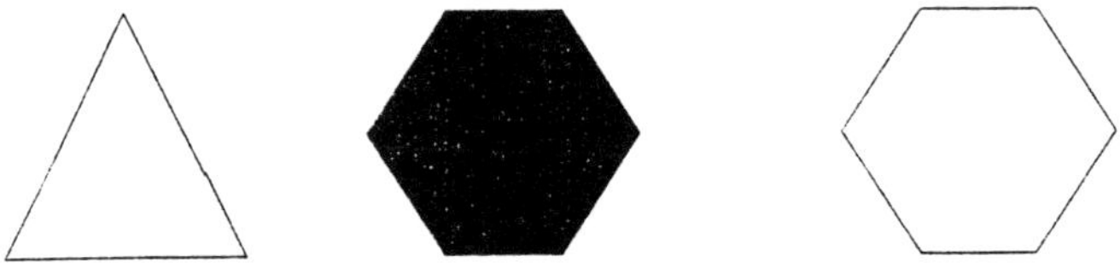

Figure 1. Some descriptions determine models

This notion of model comes from logic: a model is, in this sense, an abstract mapping of properties and relations onto individuals. This concept has an old history going back through Boole and Lowenheim and comes to its modern fruition in the work of Tarski. This sense of model is to be contrasted with what we might call *ana*logical models, which again have a long history. Some high-points in their relatively recent history are the work of Maxwell (the physicist), Craik (1943), and a contemporary example would be the work of Gentner (1981). This latter sense of model is one in which one process is analogised to another. Examples are analogies between the flow of electricity in conductors and the movement of people in a crowd through streets, or again the movement of water through plumbing.

I do not wish to claim that these two concepts of model are entirely unrelated. Both involve the notion of mappings under morphisms. But I will argue that Johnson-Laird in his "Mental Models" theory of cognition conflates these two senses. According to Johnson-Laird, mental models are a medium of representation, in particular a medium of mental representation. The alternative view is that models are important to cognitive theory because they are abstract mappings which define *what* is represented not *how* it is represented. On this latter view the main psychological question is how models are represented mentally, a question almost coextensive with the binding problem. This point appears to cause much confusion and I offer an analogy with the other central abstract object of cognitive theory, the sentence.

Sentences are abstract objects, one might think of them as the abstract objects that Chomsky rehabilitated. They have a sequential structure and they also have hierarchical structure, and we know many formal techniques for describing them. But we don't have a "mental sentences" theory of how people use language. What we have is many years of psycholinguistics which tells us about the many ways in which sentences are implemented. We know that the human comprehender, immediately after hearing an auditorily presented sentence, implements that sentence in a way which is rather close to its phonological characterization, and we know that very soon afterwards the speaker will have an implementation in which much of the surface has been lost but which still retains some implicit hierarchical structure. It is a moot point how much of the remaining sentence structure is preserved in the structures and processes which succeed sentence comprehension.

So too, I would argue, with models. Models are abstract objects which tell us what a person is trying to extract from a text, and what we need are theories about how these models are represented at various different stages of psychological processes. From this point of view, it would be extraordinary if there were just one sort of representation for models. The rest of this paper is going to describe a program of research which was designed to ask the question: "How are models represented?" particularly in the processing of simple texts which describe small sets of individuals.

III. MODELS IN WORKING MEMORY

The first requisite for such a research program is a task and the task that we have adopted, the most direct approach possible, is the task of remembering mappings. We call this the "Memory for Individuals Task" or MIT. Figure 2 shows an example trial that a subject might undergo.

The sentences would typically be read one at a time in self-paced reading task and each sentence would disappear when the signal to present the next sentence was given by the subject. Subjects read under instructions to go as fast as possible consistent with being able to remember what was in the paragraph.

There is a bishop,
There is a dentist,
The bishop is polish,
The dentist is swiss,
The bishop is sad,
The dentist is sad,
The bishop is tall,
The dentist is short.

Recall menu

tall	short
sad	happy
polish	swiss
bishop	dentist

Figure 2. An example trial

The time spent on each gives valuable information about the processes of construction of representations (see e.g.,Stenning, Shepherd & Levy 1988). Here we will mostly be concerned with evidence about the completed representations which comes from recall errors. At the bottom of Figure 2 is shown the recall menu which a subject would typically use to recall the contents of the paragraph that they had just read. The menu appears and they pick off an individual, perhaps a tall happy swiss dentist. The menu then reappears and they pick off the other individual, perhaps a short sad Polish bishop. Subjects are quite accurate in their memory for the immediately preceeding paragraph, but we use their few errors to infer the components of their representation, and it is with the analysis of the errors that I will be concerned here.

The analysis is a conventional psychological use of a confusion matrix to reveal underlying representation. The general assumption is that responses which are psychologically more similar to the stimulus will be more frequent, those less similar will be less common. The most similar response is a completely correct recall, and indeed this is the most common response, but when errors occur, they are more often responses similar to the stimulus than ones highly disparate. This assumption allows us to derive the representation's independently implemented features from observations of frequencies of responses. There are some complications in the present use of this conventional technique which arise from the large size of the confusion matrix and the resulting sparsity of data, but these are not our concern here. Interested readers are referred to Stenning, Shepherd & Levy (1988).

The picture of the representations that emerges from this data is of a distributed representation with between 12 and 20 independently represented features, which jointly represent the contents of the paragraph just read. Each feature has a truth value for any given paragraph, and the set of truth values represent what was presented. These features contain no referential terms, they simply state existential facts about the pair of individuals.

For the Vocabulary:
SHORT/TALL HAPPY/SAD SWISS/POLISH DENTIST/BISHOP

An example description: short happy Polish bishop
tall happy Swiss dentist

Feature	Value for this description
TALL HAPPY	True
SHORT BISHOP	True
HAPPY DENTIST	True
SWISS DENTIST	True
TALL SAD POLISH	False
TALL SWISS DENTIST	True
TALL HAPPY DENTIST	True
SAD SWISS BISHOP	False
SHORT HAPPY POLISH BISHOP	True
SHORT SAD POLISH DENTIST	False

BOTH SHORT or BOTH TALL	False
BOTH HAPPY or BOTH SAD	True
BOTH SWISS or BOTH POLISH	False
BOTH DENTIST or BOTH BISHOPS	False
MORE THAN ONE MATCH	False

Figure 3. The feature set for an example description

Figure 3 shows an example of the underlying representations. In the example, one feature, "There is a tall unhappy person," is true; another, "There is a short bishop," is true; but the feature, "There is no tall sad Polish person," is false as is "There a sad Swiss bishop," and so forth. The features toward the end of the list, rather than connecting properties within an individual, relate the two individuals. So for example, one feature states that there are not two short people nor is there two tall people. Another states that there are either two happy people or two sad people. These features are effectively encoding whether the individuals match or mismatch on their respective dimensions. Finally, the feature representing the number of matches simply states that there is more than one match and we know in this case that it is false.

The important properties of these representations for our theory of working memory are firstly, that the representation of binding is distributed over many features which, when they are disrupted suffer independent fates. Secondly, the features are logically related to each other. The combination of these two characteristics means that disruption generally results in inconsistent sets of values. This is easily seen by considering that there are 2^n sets of feature values but only 136 well formed pairs of people with four binary properties.

One of the consequences of this type of representation is a quite severe problem of retrieving the canonical descriptions that are demanded in the task. One can see this intuitively by looking at Figure 3 and trying to reconstruct from the state of the features what was presented. But the problem becomes much more severe when we consider that this representation might be corrupted, for example, by flipping the truth value randomly with each feature with a certain probability. This type of representation therefore requires an active inferential mechanism to impose a consistent semantic interpretation on inconsistent data. This inductive inference is at the heart of this theory of memory and sets it off from theories based on localist semantic network representations. In the latter, the resolution of inconsistency is the task of a "higher" interpreter which operates over the declarative representations.

One very natural computational solution to such problems is to use a PDP network as a retrieval mechanism. PDP systems are adept at making multiple constraint satisfaction inferences. Figure 4 shows a three layer feedforward network with the input layer at the bottom. Each input unit corresponds to one of the features in the model from Figure 3 and these are connected via a hidden layer of units to an output layer which is interpreted as a canonical description of the two individuals. One individual is represented by the four units on the left, and the other individual by the four units on the right.

The network is trained on the logical relations between the features by presenting only well-formed input vectors, and for each setting the output vectors to the represented pair of individuals. Stenning & Levy (1989) were able to show that a network trained in this fashion on the relatively few well-formed input vectors produced errors very like our human subjects when random noise was injected into the input layer, thus producing inconsistent patterns which the network had not been trained on.

Such a device gives an existence proof that this is an architecture which can readily make the sort of inferences necessary from a distributed representation to the output form in which subjects can access the material. It is important to note that the model is only intended as a simulation of *retrieval* from these distributed representations. To complete the picture accounts of the construction and storage of patterns are required. However, even this partial account is enough to illustrate the sort of explanations which a theory of working memory may be able to offer of phenomena of verbal reasoning.

This theory of working memory takes as its central goal an account of how the properties of an individual are bound to its identity. Other theories of binding in memory such as Anderson's ACT* (1983) assume binding is represented by primitive unanalysed elements of the notation, namely arcs in a semantic network. In this theory, the bindings of properties to individuals is not a simple primitive but a complex

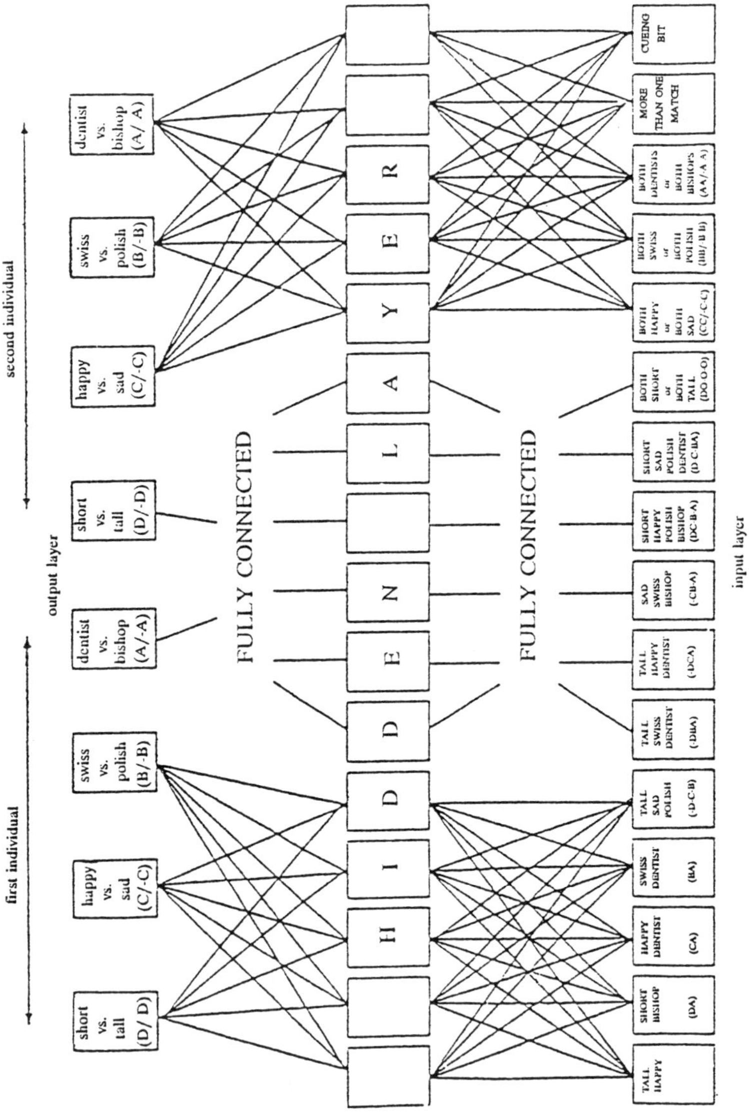

Figure 4: The PDP network

emergent. This theory offers the first real opportunity of explaining why binding is content dependent. When the theory states that there is a feature corresponding to the proposition "There is a bishop who is Polish" this means just that there is something in memory which enables the subject to discriminate Polish bishop from the other possibilities active in the situation, namely Polish dentist, Swiss bishop, and Swiss dentist. Whatever this something in memory is we doubt that it is a sentence like the one written here. Rather, we suppose that this information is represented by the recruitment of associations from general memory. One might imagine a subject in this particular example recruiting the fact that there is a person who is catholic. General knowledge of polishness, swissness, bishops and dentists is sufficient to enable an association like "catholic" to bind "polish" to "bishop."

At present the simulation only reduces the binding of four properties to two people to the less complex binding between conjuncts within features, and it does so in a way that does not make appeal to content. However, this regress is not vicious. We believe that because the theory makes binding depend on inference, and because it reduces complex structure to simple conjuncts, that this theory offers the potential of a fundamental explanation of why binding in human working memory is content dependent. The relationship between the simple PDP retrieval mechanism that we have simulated and the long-term associative knowledge underlying memory is an object of current research.

IV. REPRESENTATIONS FOR SYLLOGISTIC REASONING

How does all this relate to Mental Models theory? We agree with Johnson-Laird that people reason by considering example models—that is example mappings of properties onto individuals. But we disagree as to how the theory of the logic behind human reasoning and a theory of its implementation relate. We would argue that mental models notation is a logic, that mental models notation is equivalent to a slightly enhanced Euler's circles notation and that mental models theory does not account for content effects. The intuitive motivation behind both mental models and graphical representations is the same, and is psychologically important. What these representations express is the agglomerative strategies of human reasoning: people, unlike say natural deduction theorem provers, agglomerate the premisses they are given into a single representation.

Any representational system which performs syllogistic inference by agglomerating pairs of premisses into one representation must be capable of representing collections of types of individual drawn from the seven types defined by conjunctions of the three predicates and their negations: A & B & C ... $\tilde{A}$ & $\tilde{B}$ & $\tilde{C}$ (It happens that it is never necessary to consider the eighth type $\tilde{A}$ & $\tilde{B}$ & $\tilde{C}$). Many ways of representing such collections come to mind: tables of individuals by properties, conjuctions of sentences containing repeated variables or constants, mental models notation, and various types of "set overlap" diagram. It is worth examining the relation between mental models and one sort of diagram, Euler's circles, more closely.

Johnson-Laird (1983 p. 77ff) has made much of the fact that Euler's circle-based theories (such as Erickson's 1974) suffer from a combinatorial explosion of diagram combinations. This is because there are multiple diagrams for all but one of the four types of syllogistic premiss. For example, there are four diagrams consistent with "Some A are B" and so when combining two premisses of this form, sixteen combinations have to be considered. But it is simple to augment the notation to circumvent this problem. Figure 5 shows augmented Euler's circles representations for each of the forms of syllogistic premiss. The only difference between this and standard Euler's circles notation is that we have chosen the diagram which makes the weakest assumptions, and we have shaded the section of the diagrams corresponding to the sets of individuals which are *established to exist* by the premiss. This enables us to uniquely represent each syllogistic premiss type with just one diagram. It is of particular interest to note that the two premisses "Some A are B" and "Some A are not B," are represented by the same circles, and only differentiated by shading. "Some A are B" establishes the overlap between A and B is not empty. "Some A are not B" establishes that the part of A not overlapping with B is non empty.

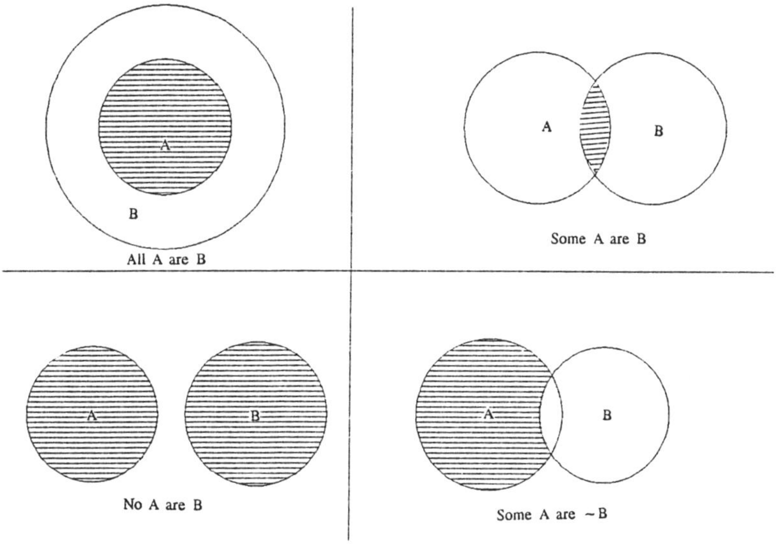

Figure 5. Euler's Circles Representation of Syllogistic Premises

Armed with this augmentation of Euler's circles notation Figure 6 shows an example syllogism, its mental models notation and the corresponding augmented Euler's circles representations. In the mental models notation, columns represent individuals. So, for example, the left hand column represents the fact that there are individuals of the type A, B and C. In Figure 6 we have shown the areas in the Euler's circles representation which correspond to the various portions of columns in the mental models notation. Notice that the parenthesized letters in the mental models notation correspond to unshaded areas in the Euler's circles diagram which reflects just the fact that it is not known whether or not the parenthesized letters are established by the premisses.

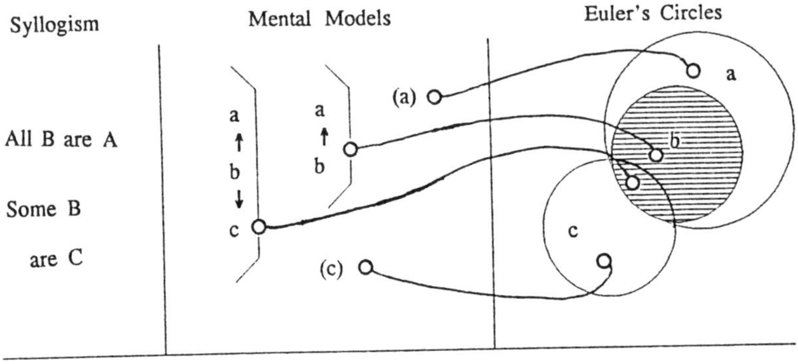

Figure 6. An example syllogism in mental models and Eulers Circle notations

We can imagine Johnson-Laird's subject having constructed this mental model's representation and now entering on testing phase with the candidate conclusion "Some A are C." An attempt to test this conclusion is just an attempt to find an arrangement which will satisfy the premises, but break the connection between A and C in all individuals. We can see that this process is modeled in the Euler's circle representation by the question whether the circle C can be "pulled away" from the circle B, thus destroying the overlap, without making one or other premiss false. We can see that just the same sorts of arguments can be made with respect of this representation that indeed this conclusion cannot be defeated. To mount a rigorous proof that the two systems are equivalent in fact demands considerably greater specification of both systems, and this is not the place. For example, the subject/predicate notation in mental models can be mirrored by a notation within the Euler system. However, this example is sufficient to illustrate how two systems apparently so disparate are much more closely related than is at first apparent.

If mental models notation is a variant of augmented Euler's circles notation what is its relation to a conventional logical calculus? It is equally clear that as far as notation is concerned it can be seen as a variant of this too. Figure 7 gives a proof of the same syllogism represented in Figure 6, in a conventional Gentzen-style natural deduction system. In Figure 7 we have shown how groups of lines of the proof correspond to particular parts of the diagram.

Representing Premiss 1	Column 1	1) ∀x(Bx → Ax)	Premiss 1
		2) Bd	Non-Empty Sets
		3) Ad	Modus Ponens 1, 2
		4) Ad & Bd	& Intro 2, 3
	Column 2	5) Bd'	Non-Empty Sets
		6) Ad'	Modus Ponens 1, 5
		7) Ad' & Bd'	& Intro 5, 6
	Column 3	8) Ad" & Bd"	Assumption
Representing Premiss 2	Column 1	9) Cd & Bd	Premiss 2
		10) Ad & Bd & Cd	& Intro 4, 9
	Column 2	11) C̃d'	Assumption
		12) Ad' & Bd' & C̃d'	& Intro 7, 11
		??)	
		??) Ad & Cd	& Elim 10
		??) ∃x(Ax & Cx)	∃x Intro 13

Figure 7: The beginning of a natural deduction proof of the syllogism from Figure 6, showing correspondences to mental model notation.

This proof is certainly odd by the standards that we are accustomed to from logic textbooks. It is wildly redundant. Many assumptions are made but never used. But this redundancy is the whole lesson. Qua logic, all that logic does, is to define a set of deductive relations. When we come to ask how logic could be implemented as a device for finding conclusions or for constructing proofs of identified conclusions, then the computational characteristics of the implementation will determine entirely what is easy and what is difficult. If human computational architecture makes a representation of rich models easy but it makes simultaneous consideration of several such models rather difficult, then a process like the one laid out in this proof may well prove a sensible implementation for that device. Of course it is very different from the implementations that we would use on a conventional computer but then that was the point that we started with, human beings aren't a conventional computer and they don't implement logic in the way that we are accustomed to. Pencil and paper allow us to learn proof heuristics which are utterly unattainable to our customary working memory mounted "mental logic." The "mental" in mental logic here is to be interpreted by analogy with the "mental" in mental arithmetic. Natural deduction systems are about as accessible to our working memory as are the long division methods we learn at school, but that is not to say that logic or even long division cannot be performed in working memory given the right heuristics which exploit its strengths and avoid its weaknesses.

V. DISTRIBUTED BINDINGS AND SYLLOGISTIC REASONING

Our consideration of representational systems identifies the agglomerative nature of both mental models and Euler's circles as their psychologically important characteristic. We agree with Johnson-Laird that people approach syllogisms by constructing representations of models (in the logician's sense) and then testing them against candidate conclusions. When several such constructions and tests must be carried out in series, the problem becomes quite hard. Where does our theory of working memory fit in with this picture of syllogistic reasoning? What does a working memory that depends on distributed representations from which material is retrieved by inductive constraint satisfaction explain about syllogistic reasoning performance?

This type of working memory explains why the human strategy is to agglomerate the premisses from which an inference is to be drawn. Such a memory has a relatively high capacity for a complete and determinate mapping of properties onto individuals, but without augmentation, it cannot represent incomplete ones, because presenting a constraint satisfaction network with incomplete input does not lead to incomplete output. Furthermore, this type of memory explains why human beings find it difficult to entertain more than one logical model at a time. PDP systems performing parallel constraint satisfaction inferences are radically serial. The whole network can perform only one such inference at a time and to perform another requires a resetting of its input units followed by a series of settling cycles.

The theory also helps to explain one sort of content sensitivity of human syllogistic reasoning. If the working memory representations of models depend on the associative base for the implementation of their low-level features then human working memory from models will be dependent on content. If reasoning is dependent upon memory for models then reasoning will also be dependent on content. It is also clear that entertaining a sequence of models which consist of the same content rearranged in different permutations will lead to enhanced interference on multi-model problems. This is of course by no means the only type of content dependence in human reasoning, but it seems to be the dominant one when undergraduate subjects perform syllogistic reasoning with the content of the type that Johnson-Laird uses.

There is, of course, more to syllogistic reasoning than this sort of working memory for models. Subjects need to select which models to represent, to formulate candidate conclusions, to test them against the models constructed, and to keep track of all these goings on. For present purposes we assume that these processes go on as described in the mental models theory working over models which are held as the output of the network we describe.

Of course a great deal of work remains to be done to show that these are the memory structures that human beings are using during syllogistic reasoning, but I think it is not an unreasonable claim to say that the memory for individuals task models a simpler component which is the basis for syllogism solution.

Our general conclusion is that if this view of models is right, and their importance for cognitive theory is that they are abstract objects which define what is represented, then the hard psychological work is to show how implementations lead to the resource limitations that we observe. We have offered a sketch of one human implementation of working memory for the logician's models which explains why some observed limitations are as they are. On this view, the problem is not one of inventing natural logics or analogics which will explain human reasoning but simply to explain how logic can be implemented on a device which has a memory architecture so radically different from conventional computers. It is perhaps as well to remember, at this point, that human working memory did not develop this architecture for the purpose of solving syllogisms. The mechanism is primarily adapted to the *inductive* task of deriving a unique best-fitting model from a rich array of input, context and general knowledge. This is not a deductive task, let alone one performed according to classical logic. Using this model-deriving mechanism to underpin syllogistic reasoning is a little like taking a computer system to a lecture because its printer contains the only available notepaper.

REFERENCES

Craik, K. [1943] *The Nature of Explanation*. Cambridge: Cambridge University Press.

Erickson, J. R. [1974] A set analysis theory of behaviour in formal syllogistic reasoning tasks. In Solso, R. (ed.) *Loyola Symposium on Cognition*, Volume 2. Hillsdale, N.J.: Lawrence Erlbaum Associates.

Gentner, D. and Stevens, A. L. (eds.) [1981] *Mental Models*. Hillsdale, N.J.: Lawrence Erlbaum Associates.

Johnson-Laird, P. N. and Steedman, M. J. [1978] The psychology of syllogisms. *Cognitive Psychology*, 10, 64-99.

Johnson-Laird, P. N. [1986] Conditionals and mental models. Chapter 3 in Traugott, E. C., ter Meulen, A., Reilly, J.S. and Ferguson, C. A. (eds.) *On Conditionals*, pp 55-76. Cambridge: Cambridge University Press.

Stenning, K. [1975] *Understanding english articles and quantifiers*. University Microfilms.

Stenning, K. [1978] Anaphora as an approach to pragmatics. In Halle, M., Bresnan, J. and Miller, G. A. (eds.) *Linguistic Theory and Psychological Reality*. Cambridge, Mass.: MIT Press.

Stenning, K. and Levy, J. [1988] Knowledge-rich solutions to the 'binding problem': some human computational mechanisms. *Knowledge Based Systems*, 1.

Stenning, K., Shepherd, M. and Levy, J. [1988] On the construction of representations for individuals from descriptions in text. *Language and Cognitive Processes*, 2, 129-164.

Wason, P. C. and Johnson-Laird, P. N. [1988] *Psychology of Reasoning: structure and content*. London: Batsford.

CHAPTER 6

ON THE STUDY OF LINGUISTIC PERFORMANCE*

Victor Sánchez de Zavala
San Sebastian, Spain

I surmise that a Colloquium such as the present one is not a bad place to take with quite a bit of salt the very notion of cognitive science—an expression I always use with some trepidation. Not only, of course, because of the well known fact that the form of words "so-and-so science" is used when there is in fact no real science covering the domain we refer to with such a "so-and-so"; other reasons suggest that some scepticism would be in order.

For one thing, under this banner we find such a number of approaches, theories and methodological criteria that we may be inclined to think of it as a mere aggregate of research programs without any other rhyme than that conferred by that very name—a hyperprogram propelled by the not very reasonable assumption that what is good for somebody's (*cognitive*) research is good for yours too. It is true that several of the disciplines so joined—or linked—have been longing for years for a coordination of efforts (so, e.g., linguists and logicians used to lament each other's ignorance of their fields and technical resources) and that in many cases where this situation was put to an end considerable progress, at least technical progress, was made. But advances achieved in a number of tasks that point to a common goal do not necessarily mean getting nearer to that goal—witness the advances in medicine (resulting in an increased population), production technology and management (more free time), and means of enjoying mobility at will (multiplication of individual vehicles), which combined have originated the growing menace of our big cities to an individual's freedom, quality of life and even mobility.

More seriously, it could be pointed out that even though learning about ideas and intellectual advances in fields different from one's own is perhaps the best source we have of new conceptions and methods, it would not be exceedingly rare, I think, that the team spirit, when pushed to the point of forming teams of teams (and so favouring the emergence of wonderfully general consensa), result sometimes in lines of research that instead of viewing from nowhere go in effect nowhere.[1] (Admittedly, there is no general system to avoid, either individually or collectively, the intrusion of misconception or error in our thoughts, and intercourse of many minds may be conducive to a sharp critical examination of notions and assumption. I only tried to suggest that the sound of many voices singing what I sing **may** obscure the fact—if it is a fact—that we are all out of tune.)

Again, in a similar way to the dictum about history, that who ignores it is condemned to repeat it, it seems to me that our usual regretting that older scientists and thinkers would not notice the assumptions of their *Zeitgeist* (let us suppose that we do not just don't care for what they thought and so rediscover it once and again) makes a strange but very common bedfellow with unquestioningly assuming the general vantage point of our own time. Let me give you an example. The very name "cognitive studies" silently merges two different issues: first the nature and characteristics of "higher" mental functioning both of humans and animals; and on the other hand what property is that of being cognizant, what is necessary—and perhaps sufficient—to know something. We could even distinguish a third issue covered by this term: what conditions and constraints must be complied with by any actual process or operation to be properly called *cognitive*. And I submit that as long as all these issues are not structured both internally and to each other in the most articulated way, risks of carefully carrying out misguided or irrelevant research are exceedingly high.

(An obvious reply would say, in a rough and ready way, that these three standpoints are respectively assumed by neurophysiology and general cognitive psychology, by "naturalized epistemology" or perhaps psychosemantics, and by AI plus empirical studies of cognitive performance, such as those carried out in psycholinguistics and in experimental —and possibly "ecological"—cognitive psychology; all of them being reflected upon—optimists might say "guided"—by philosophy of language and philosophy of knowledge, and supplied of technical means by the formal disciplines, especially logics and computability theory. Finally—this reply would continue—the so much advertised interdisciplinary nature of cognitive studies would effect, in an explicit way, the merger I so disparagingly pointed out, taking also care of the urged structuration. I agree—but on the label of disparagement—as to the intention of effecting a merger. My point, though, is that instead of simply working from the assumption of such a merger as a task feasible almost by itself, by simply pooling separate bodies of knowledge and technical skills, we should make a previous effort to unearth a final target, an overall goal towards whose attainment all these endeavours would be organized and structured. I emphatically do not mean **the** target, a goal that once discovered by a lucky cognitive scientist or philosopher would be valid for guidance of all workers in the field. On the contrary, I believe that **each** of them should try to identify it, even as a subsidiary goal to a more vaguely delineated general goal of all cognitive studies, as thoroughly as he or she will be able to. And I would claim that the multiplication of such carefully thought out "vanishing points" of research do not subtract from, but adds to the chance of getting all of us on the right track—not least because of the healthy scepticism about one's own scientific "ideology" that it would instill into everyone.)

As to myself, I'll try to take the prescribed medicine (or a first sip of it) in the present paper. In section 1 ("A proposal for "cognitive science" and a specification of it") I will describe what, by my lights, is the final target that cognitive studies should propose. Section 2 ("Current situation in Linguistic performance theory") will be devoted, first to specify such proposed target as to the field where I could not outragingly presume having any competence, and second to rapidly review the recent evolution and present situation of work on it.[2] Specifically, in this second part of section 2, I will first (paragraph 2.1) comment briefly on the comparatively scant work done on productive speech performance, and then (paragraph 2.2) I will describe the main lines of research on receptive speech performance, singling finally (§2.2.2.3.2) the significant new theory of relevance put forward by D. Sperber and D. Wilson. The final section of the paper (section 3: "Some issues regarding research programs on linguistic performance") will summarize the main shortcomings of the reviewed work and advance a few preliminary suggestions regarding possible ways of overcoming these difficulties. (In the Appendix I will present in more detail and critically examine the theory of relevance, to my mind the most significant current line of research in the field.)

I. A PROPOSAL FOR "COGNITIVE SCIENCE" AND A SPECIFICATION OF IT

Two main goals of research, I think, can be found in our "science." Research guided by one of them aims at understanding some or other aspect of cognitive activity. I do not mean only, not even principally its different types in correspondence with sensory modalities or, in general, with the "modules"[3] concerned or mainly concerned; I rather have in mind those aspects of it that are determined (for obvious reasons of limited expertise, almost always within a "module" or even a "submodule") either by a distinction of levels of abstraction in the study of this activity or by the degree of formalization chosen for the research. Classical examples are the study of syntax (of a language, of natural languages in general) or of its acquisition, investigations on the dynamics of neurons and/or neural assemblies, or philosophical research on so-called content, either of mental states or of (specific types of) expressions.

Cognitive research, though, may aim "higher up," at cognitive activity itself; even though most times such an endeavour is again limited within a mental "module" or "submodule" (a limitation even more necessary now, *caeteris paribus*, given the integrated nature of its subject). The study, say, of linguistic performance, or of "practical reasoning," are, I think, reasonable approximations to such a genre of research, first steps in that direction.

Now it seems to be obvious that this second type of study, on the one hand, can only be carried out fruitfully after some (probably, quite a bit of) progress has been made along the other line of research; and also that its avowed target is just the best current approximation to the final goal of cognitive research in general, i.e., to fully understand the conditions and properties required (in its "bearers" and their environment) for cognitive activity (the phenomenon "without which not" in its more radical sense: nothing actual—or instantiated, if you are a Platonist—to study, no one able to study anything).

As a matter of fact, I consider this goal exceedingly limited to be really interesting. The nature of cognitive activity itself (the processes, at their various levels of description, that implement it, their effect on the intellectual resources of the system—let us say, man—that carries it out, and similar matters) should certainly be included in a not too shortsighted view of our "science." But even this goal seems to be somewhat unambitious; for cognitive resources and activities are—no surprise meant—something artificially isolated from the total fabric and functioning such a system, man, where many other aspects intermingle in an intricate but searchable manner with them (as testified, if there was any need to do it, by current work on cognitive aspects of emotion). So, not because of any chimerical ambition to achieve a total science of man (and the Universe), but out of an anticipation of a possible need to effect a mesh between cognitive theories and studies about other facets of man and his activity (and so be able to give satisfactory answers to "mortal questions"), I think necessary to take into account, in some way or other, such a future *mesh* and start aiming at cognitive activity itself as embedded in an aggregation/system of many other activities.

Let us return to the observations I made about the second type of study described a couple of paragraphs back. While I take it that the second point (cognitive activity itself may be considered the final target aimed by cognitive science) will secure an almost unanimous support, due to its being on the verge of triviality, the first one (detailed research of specific cognitive phenomena as a precondition to any attempt to investigate cognitive activity in general), while begin not less certain to me, allows a wide variety of opinions as to the precise amount of knowledge that must be marshalled (by the scientific and philosophical community) in order to fruitfully launch the *holistic* type of investigations we are now concerned with. As to myself, I would venture the suggestion that exploration of as yet scientifically uncharted frontiers could be, if carried out with due caution, certainly worthwhile; not only for philosophical illumination but even in order to help give a direction to future scientific research.

My suggestion is then to initiate (or continue already initiated) studies of cognitive activities, with a view to eventually reaching a full theory of cognizant subjects to their natural environment. My immediate goals are though, understandably I think, somewhat more modest: I propose simply to explore the full range of **linguistic performance** from the standpoint of its relationship to other "higher" activities mainly but *not exclusively cognitive* (in a narrow sense of this word), i.e., without unduly restriction to what is usually called "thinking." All other research goals currently posited in studies of language (be them philosophical, strictly linguistic or psychological, to name the main disciplines interested on it) are, to my mind and no matter how great the need to pursue them for reasons of feasibility or just expediency, no more than preliminary and by their very nature minor, subordinate to that one.

II. CURRENT SITUATION IN LINGUISTIC PERFORMANCE THEORY

I will concern myself exclusively with spoken language, since writing and reading, no matter their undeniable cognitive specificity, are certainly secondary processes built upon oral—or, for the deaf, signed—speech, and should be studied at a later moment. As regards (oral-auditive)[4] linguistic production and reception, I will deal again only with studies of primary phenomena, i.e., those occurring in face-to-face interaction (if interaction there is at all), albeit in most cases they also occur in other situations of linguistic communication. This policy, on the other hand, allows exploitation of most of existing literature on (non-sociological aspects of) language use, so that little if anything is lost as an effect of the restriction imposed. (There should be no commitment to it, of course, were it to happen otherwise).

Not surprisingly, linguistic reception (variously named "perception," "comprehension," "understanding..." of language) is much more studied than production. The manifest, public nature of the *terminus a quo* of reception, open to as much examination as desired, and the fact that it is comparatively easy to follow up the "decoding" process, at least up to and including parsing of the sentence, and even to ascertain (perhaps only in a gross manner) whether its *terminus ad quem*, the message being understood, is reached and when it so happens, contrast sharply with the total unknowability—or so it seems—of the *terminus a quo* of linguistic production, that apparently can only be investigated just through its final product or *terminus ad quem*, the sentence or sentences uttered. As a result of this difference[5] there is, concerning production, *(a)* a real dearth of data and theories, *(b)* a suspect extremely close similarity in almost every detail (but to specialist's eyes, of course) between the descriptions and explanations offered by different researchers—if we ignore "analysts of discourse," to which I will directly refer. (I called the similarity suspect since I believe it betrays no more than the paucity of evidence on which to build any theory of the process, if we ignore "low-level" data, i.e., those about command and control of speech organs, that hardly can be considered cognitive.)

2.1 Linguistic Production

As just hinted at, results secured by either psycholinguists who study more or less spontaneous speech (mainly through analyses of *slips of the tongue* or pauses and changes in rhythm) or who in some way or other manage to have their subjects utter previously chosen expressions (see e.g., Fromkin [1973] and [1977-80], chs. 9 and 10 of Cooper and Walker [1978], Osgood [1979], Butterworth [1979] and [1982], Cooper and Paccia-Cooper [1980], Garrett [1981], Cutler [1982], Lapointe [1983], Levelt [1983], Motley [1985]) are meager as concerns cognitive processes in speech production that are not narrowly linguistic. It is very rare to find them hypothesizing anything deeper than an initial level or phase for representation of the gist of the **intended message** as an overall organizational schema of the sentence plus a few selected key lexical items or perhaps *abstract* semantic elements—a schema that would be later specified, under more or less continuous monitoring, down to the needed articulatory detail.[6] (For an early plea in favour of a non-serial but interactive process of sentence formation, that parallels empirically warranted hypotheses about sentence reception—as stated below—see Bock [1981].) In a few cases the need for a previous fixing of illocutionary force is acknowledged, but usually in a cursory way—see, e.g., Bach and Harnish (1979: ch. 11), to which I will return. Apart from some textbooks and manuals (notably Clark and Clark [1977], chs. 6–7) where most relevant research in analysis of discourse, speech acts, degrees of informativeness and saliency (*given* versus *new*, etc.) is organized and summarized in a series of decision steps that the speaker is assumed to take, the exceptions come from computer simulations of the speaker (*vid.*, Appelt [1984]), where we find, first, **planning** of the illocutionary and locutionary means to achieve desired results on the hearer (a **model** of which is also provided), and second, a detailed specification of resources and operations to be effected until the complete sentence is constructed.[7]

I wrote a moment ago that workers in *discourse analysis* differ from most other researchers on speech production in that they (discourse analysts) **do** consider cognitive processes in speakers. It should be added now that they offer an alternative and—it seems to me—more flexible conceptualization to the (Austin-Searle, it could be dubbed) theory of speech acts. Unfortunately, though, the reason why their pragmatic theory has such flexibility (their intellectual heritage, through Schulz, of Husserlian phenomenology)[8] is also responsible for their not providing much help towards the goal I advanced in previous paragraphs. For they pay so much attention to these processes, and are so anxious to be faithful to their true nature that highly detailed descriptions of (presumed) individual goals, means, communicational "moves" and strategies substitute for a theory of the linguistic intercourse they study. (This does not mean that their work is worthless; it only means that their sometimes shrewd observations and concepts must be evaluated and interpreted by means of resources entirely foreign to their standpoint.)[9]

2.2 Linguistic Reception

Work in this area can be usefully separated in three main classes. First, studies devoted to the procedures and mechanisms that enable the hearer to divide up the continuous sound flow of speech in discrete units (phonemes, morphemes, words and even longer lexical items) and identify them thanks to some process or other of access to lexical entries. Other research either examines methods presumably used to find out the structural characteristics of the expressions heard (up to and including whole sentences) or advances models of such putative methods, the idea being that by their use the hearer is able (or maybe just helped) to get at the meaning of the message. And understandably the proposals here are heavily dependent on commitments to particular linguistic theories.[10] Finally, for some psychologists, linguists and philosophers it is immaterial, at the present situation of research, what particular syntactic or lexical operations are carried out (and even more so, what detailed computational tasks make feasible such operations) in understanding a sentence, or a discourse; the main theoretical task in this field lies for them in delineating the basic resources and general method used by hearers to understand linguistic messages; and so they advance "overhead" theories independent of theories about lexical access, the nature of syntactic structures or parsing procedures.

I will only deal with the last two types of studies—the first one seems clearly to be by far less concerned with cognitive, especially general cognitive, operations. In fact, only the third type will be at all reviewed (however cursorily, for reasons of space), since its general stance is most germane to that presented in this paper. And it should be noticed that while centering on sentence understanding, i.e., on linguistic receptive performance, is the rule (excepting perhaps approaches from a strong psychological standpoint), a theory of this type usually includes also some sketch of productive performance.

2.2.1 Studies About Parsing and Understanding Language

As it was just said, investigations dealing with putative methods and processes of using the syntactic structure(s) of the sentence to recover its semantic import show great differences according to syntactic approaches accepted. And since the field was until recently (by the final years of the seventies) dominated almost to the exclusion of any other viewpoint by generative transformational grammar, mainly in its *extended standard* variety, the seemingly very first proposals about "strategies" of parsing by the hearer (Bever [1968-70] and especially Kimball [1973] and [1975])[11] were followed within this central line of linguistic research by a whole series of investigations (see, e.g.,Cooper and Walker [1978]),that perhaps is best represented by the so-called *sausage machine* (Frazier and Dean Fodor [1978],Dean Fodor and Frazier [1980]).

On the other hand, some researchers working within a strictly experimental setting (Marslen-Wilson [1975] and [1976]; cp., Swinney [1981]) show convincingly that the process of "decoding" spoken sentences cannot be serial (as it was tacitly assumed by the previously mentioned investigations); phonological, syntactic and semantic structures identification (reconstruction, interpretation) interact in such a tight way that they can be considered parallel or almost parallel decoding processes. As a matter of fact, they prove conclusively that the hypothesis that parsing-interpretation waits until the sentence is completed, or is at least effected clause by clause (as it was natural to think for psycholinguistics accepting generative transformational grammar), is false: multiple (parallel) "decoding" processes proceed almost continuously (perhaps even word by word) from the first word or first couple of words down to the end of the sentence (in a "left to right" manner, as it is usually said). This result is mostly ignored for several years by workers on parsing strategies, but finally wins general recognition; and as we will see, both supports some new grammatical proposals that put at the forefront linguistic performance and must be considered an effective constraint imposed on any theory of language interpretation that seeks "psychological validity."

Before any of these studies took place, creators of "high-level" programming languages had to include parsing methods for their artificial languages (where no transformations occurred); and some workers on computer languages parsing attempted to extend directly their methods to a theory of human language parsing (Woods [1970],Kaplan [1972] and [1974];see also Wanner [1980]).These technical developments eventually influenced the thinking of linguists about how the syntactic structure of a sentence (of a natural language) is recovered by the hearer, giving rise both to attempts to develop grammars with a modicum of transformations (Lakoff and Thompson [1975], Bresnan [1975-76], and many others afterwards) and to computational models (programs) of parsing devised for natural languages under the usual assumption of their having a standard transformational grammar (Marcus [1977] and [1979],and later refinements; a general review of the field is Berwick and Weinberg [1984]).

The different lines of research that emerged as just indicated (which of course are continued by many investigators and argue almost unceasingly with each other) are not, however, directly relevant for my present purpose, since they are not less *encapsulated* than the presumed phenomena they study; they are exclusively concerned with linguistic processing by the human hearer, not with cognitive processes in a broad sense that include as a proper part what we ordinarily call "thinking"(and whose interaction with language performance is just the focus of interest here). So, this rapid survey must proceed to the other approach mentioned above.

2.2.2 Theories of Speech Interpretation

Again it is possible to distinguish in this area several lines of research.[12] When these theories arise two distinct but intermingling traditions can account for most if not all of the studies; philosophers and linguists' approaches, albeit from very different ascent, blend initially (and perhaps also later) in an almost inextricable way. In addition one must pay attention to psychologists' investigations—some of which blend in turn with concerns peculiar to either computer scientists and/or literary scholars.

2.2.2.1 Philosophical approaches. As to origin, if we momentarily ignore where Davidson's conception of semantics comes from (i.e., from the ban to meaning due to Quine), the source of everyone's theory of interpretation, in the sense here intended, is Austin's pragmatics; and not so remotely, but also looming large for most proposals I am now dealing with, we find Grice's ideas on meaning (1957)[13] and conversational constraints on communication (1967). We will see later how the conceptions of some philosophers and linguists seem to converge again; but it must be recalled now how Grice's theory of conversation contrasts *what is said*, that apparently would be immediately recovered by anyone knowing the language used by the speaker who would distinctly perceive the words he used, and *what is (conversationally) implicated* by the uttered words in the particular situation of speech. The latter is, then, a "part" of the intended message that, not being logically entailed by what is said, can only be recovered through (non-conclusive) inference based on extra-linguistic knowledge (among other things, about some principles and maxims that would constrain conversational interaction).

While trying to justify his circuitous semantics, Davidson fell upon the idea that if his truth-based theory of meaning was right, then it would satisfactorily account for comprehension of sentences of another person's language[14] (i.e., to his mind, it should be possible to provide a translation *salva veritate* to one's own language). This notion, combined with another one that seems native to him, that to understand human language it is necessary to have a theory of human action,[15] supported what—disregarding several previous hints—may be considered his initial account of the process of understanding a sentence (Davidson [1971-74]), as well as other more developed versions (see Davidson [1973a], [1973b] and [1974]), where an assumption of veracity of the speaker (consistency between his beliefs and what he means by the uttered words) plus a somewhat weakened Quinean "principle of charity" (the assumption that his beliefs are not very different from those of the hearer)[16] were thought sufficient to provide the desired truth-based interpretation.

As it is well-known, building explicitly on Grice's theory of meaning and on several Davidson papers, Lewis (1974) advances a similar but more detailed theory of how can be solved (of course, in a non-Quinean way) the critical Quinean problem of radical translation. Like Davidson, he assumes a recursive specification of the sentences of the language (in fact, of their truth-conditions). And again he thinks that knowledge of such a systematic way of specifying the semantics of sentences is not at all enough to understand speech; only through the assumption of several principles that would presumably constrain the speaker's attitudes and behaviour (including his linguistic behaviour), as well as the relations between them, it would be possible for anyone to find out what the speaker meant by saying what he said.

(He claims that several principles must be posited beyond those accepted by Davidson. Specifically, in addition to a supposition of semantic "**generativity**," he suggests *in toto* the following assumptions: *a)* the speaker's attitudes—beliefs, desires etc.—being **rationally** understandable given his complete physical state—and, by "**charity**," not importantly different from the ones the hearer would have were himself in the speaker's situation—should be able to be approximately guessed and attributed to him; *b)* this state of the speaker would tend to be **manifested** in his dispositions to speak; *c)* the speaker would abide by the conventions of his language and will expect others to act in like manner, so that the objects of the beliefs and desires accurately attributed to him—by rational and charitable guessing—would satisfy the **truth** conditions of the sentences he uses; and *d)* the attitudes attributed and those that the speaker would manifest in his language would—taken into account their respective contexts of situation—mirror each other by virtue of the **triangular** relationship between both classes of beliefs, desires etc.—with their contexts—and the propositions expressed by the used sentences.)

Apparently this type of theorizing soon loses much of its appeal; perhaps as a result of the abrupt raising of the standards of explicitness and logical rigour effected around 1970 by Montague's semantics, on one side, and, on the other, because of the rise of a new (Oxonian) philosophical semantics—just around Dummett's ideas, which we saw reacted to Davidson's—and the development in the hands of Searle, at Cambridge, of Austinian pragmatical theory. In any case, the ideas about linguistic activity that soon became more common among philosophers (at least for those tending to sociological approaches) were stamped by the theory of speech acts and in a few cases by the theory of conversation (it should be recalled that in his essay [1974] Searle himself partially reinterprets his own theory by means of Grice's).

2.2.2.2 Psychological Approaches. Almost immediately after generative transformational grammar was created, some psychologists were interested in knowing if and how differences in grammatical complexity were perceived by language users and, more generally, by which means hearers (and readers) come to understand different types of sentences in different situation. Early studies in this direction (Wason [1959] and [1961]) multiplied rapidly and were followed by many others, so that fairly soon there was a growing literature on this and related subjects (such as, for instance, the psychology of reasoning).

These developments, due to psychologists trained in experimental techniques and foreign to any notion of building a formal model, linguistic or otherwise, of mental processes, resulted in some cases[17] in the line of investigation into "strategies" of perception of sentence structure I mentioned previously. But most times they remained within the tradition of eschewing as much as possible syntactic issues and focusing in perceptual and lexical facts to explain the evidence. And this tendency was strengthened when researchers from so-called ecological psychology joined in to these studies (see, e.g., Bransford and McCarrell [1972]).

But, on the other hand, unexpected progresses in AI allowed writing programs that would (parse and) "understand" sentences of a natural language (Winograd [1972]), a fact that stimulated research on the simulation of linguistic communication, and leaded rapidly to a collaborative effort between cognitive psychologists and computation theorists (that has continued to this day). And this mutual influence, that rapidly generated work as significant for cognitive studies as, e.g., Minsky (1974), helped to develop a whole line of research into models of both sentence and more extended texts understanding that rapidly incorporated ideas developed by "econological psychologists," and even (partially) blended with literary studies about the structure of narrative texts. But since all this work aims in effect to develop models of knowledge representation (at best, then, semantic models) and of its exploitation for understanding texts, and most times with a strong bias towards design of technical systems able to "understand" information (no matter their similarity, if any, to human ways to carry out similar tasks), its significance for the kind of research focused in the present paper can only be secondary—as provider (at best) of some constraints on mechanisms of information storage and retrieval.[18]

2.2.2.3 Pragmatic Approaches. I have already mentioned the two major sources of thinking about language understanding from which derives present thinking by linguists on this subject: Searle's theory of speech acts and Grice's ideas on meaning and on conversational constraints of such understanding. (It is interesting to recall the fact that both writers—like Austin was—are philosophers, not professional linguists.

And it could also be noticed that all writers I will deal with in this section but Dascal seem to be blissfully unaware of—or just unconcerned by—Davidson's and Lewis's suggestions about interpretation of sentences.)

2.2.2.3.1 Initial Theories. A first effort to go beyond conventional (i.e., Searle's) description of what must be fulfilled in order for utterances being adequately understood, in other words, beyond the concept of preconditions for speech acts to be achieved and taken up, was that of Sadock (1972). This attempt, though, was carried out within a framework of semantization of pragmatics (ultimately originating from Ross's *performative hypothesis*) that, no matter the ingenuity of some of its proposals, pushed the whole enterprise outside the domain we are interested in—a host of posited and derived sentence structures substituted for an account based on interactional, mental and/or neurophisiological processes.

A significant step towards an adequate theory of linguistic performance is taken in Bach and Harnish (1979). These investigators[19] try to amend and expand Austin's theory in order to build a model of linguistic communication in a strict sense,[20] and to that end they re-examine and develop Grice's theory of meaning, where the audience must execute an inference—there is no other general way to do it—in order to find out which was the *reflexive intention* (see last note) of the speaker. This inferential task is pursuant, they think, to a number of assumptions that should be shared or "mutual" if communication is to succeed; some of them are specific to the particular communicative act being effected (*mutual contextual beliefs*),[21] while others are general for linguistic communication—independent of specific situations and types of them. And they try to identify the latter[22] in order to give an account of such communication in terms of inferences made by the hearer—to present a model of linguistic (receptive) performance. (They do not at all attempt any theory of psychological or subpsychological implementation of their inferential model.)

A careful review of previous research by psychologists and philosophers on the principles and possible general steps followed by the hearer to interpret sentences (and even longer stretches of discourse) can be found in Dascal (1983),[23] where in addition to a discussion of a number of conceptual matters (e.g., the very concept of pragmatics and its distinction from the study of linguistic performance) several interesting issues are raised. Firstly, Dascal claims that *non-communicative* uses of language (where linguistic performance is, of course, not receptive but *productive*) are on a par with its communicative uses and should in principle be investigated as much as these. He also emphasizes that no matter whether the correct process of interpretation of an utterance should be "direct" or "indirect" (in other words, whether the "*speaker meaning*" is, as regards content, identical or not to the "*utterance meaning*" or meaning—including illocutionary force—associated to the utterance by virtue of the conventions of the language used) it must include context-related, pragmatic steps—a fact that because of the unavoidability of resorting to a variety of hypotheses regarding mental state, intentions, etc., of the speaker results in such inferential process being not only of a heuristic, but of an *abductive* nature. Finally (in §§2.4,2.5,2.8 and 3.2) he makes some suggestions as to the specific steps presumably followed by a hearer in order to arrive at a satisfactory hypothesis about the "speaker meaning."

> (Most of these steps would be only needed when an "indirect" interpretation would be apposite, i.e.,when the issue about whether or not a computed *utterance meaning* could be plausibly considered as what the speaker in fact meant to express by his/her utterance —in other words, as identical in content to the *speaker meaning*—receives a negative answer. But in order just to decide the issue the hearer's beliefs about the discourse situation and his/her "folk theories" about people's communication behaviour, about psychology of people in general and about the speaker's psychology would play an important role.)

2.2.2.3.2 Sperber and Wilson's Theory of Relevance. In fact, this is not primarily a theory of speech interpretation, but a general theory of human communication. Nevertheless, it (understandably) focusses on communication by means of language, and its main concern is with the interpretation of sentences (even though it assigns to the speaker tasks "symmetrical" to those that the hearer presumably has to carry out in order to figure out such an interpretation). The theory (that will be fuller presented, and criticized, in the Appendix) may be best seen as supported by three basic theoretical constructions and a methodological ideal: a criticism of Grice's account of conversation, a theory of man's cognitive system, a theory of human communication and the ideal of a science with a minimum of axioms or principles, accounting for a rationally transparent object and expounded in a maximally formal way.

As regards Grice's theories of meaning and conversation, Sperber and Wilson take an initial stance quite similar to that adopted by Bach and Harnish,[24] in that they object to the hypothesized *mutual knowledge* (a rejection that they carry to the bitter end) and they assume that the interpretation of an utterance is *always* reached through an inferential process—so that there would be no processing essential difference between what Grice considered "said" and what he deemed "implicated."

Their methodological ideal, though, drives them to ask for either a principled positing of principles of conversation (i.e., of interpretation of utterances in conversation) or the replacement of all of them by an absolute minimum—just a single principle. And in their theory they take this second tack, accepting only a *Principle of Relevance* (in previous versions, of *maximum* or *optimum relevance*), with a content that does not seem very different from Grice's Principle or Maxim of Relation ("Be relevant!") but that in fact it is. For, (*a*) it is not a maxim, but a principle always adhered to by the speaker and the hearer, and so counted upon by both, (*b*) the notion of *relevance* is changed from its usual content, so that it now varies in proportion to the amount of (non-trivial) consequences derivable, by rigourous deduction, from given premisses (amongst them the propositional content of the sentence or sentences uttered), and (*c*) the processing cost of this derivation or inference and the time needed to effect it must be kept at a minimum (while preserving a sufficiently high relevance).

The theory of the (human) cognitive system Wilson and Sperber propound may be reduced to a couple of main ideas. First, that such a systems' function is necessarily that of trying to get a maximum of information with a minimum of processing and time costs. (This is, given their theory of communication—that will be presented in simplified form directly—the rationale for aspect (*c*) of the Principle of Relevance.) And second, that such information (plus possible innate information) forms a huge body of (mostly tacit) assumptions of propositional form—whose content is sometimes called our *encyclopedia knowledge*.

Finally, Sperber and Wilson's theory of human communication is predicated on two assumptions, one unstated[25] and a second one clearly stated. The first one is, that human communication is only communication of information, the other, that what communication achieves is a change in some of the receiver's assumptions forming his/her encyclopedic knowledge, not a bare addition of the content of the message to such a body.

The theory can be summarized in the following way. First of all, the hearer tries to get at the message intended by the speaker by assuming that he or she complied with the Principle of Relevance, i.e., that (s)he meant to convey a message having sufficient relevance for the hearer (or at least for the addressee), and also that the message is recoverable with a minimum of processing time and cost. Second, the hearer extracts the propositional content(s) of the sentence(s) included in the utterance heard and adds to them the contents of other assumptions taken from his/her own encyclopedic knowledge (and possibly others built at the spur of the moment) so that from this specially formed [*doxastic*] *context* it will be possible to derive, in a strictly deductive fashion, a consequence —an interpretation of the message—that preserves his/her initial assumption, the stance of relevance. And finally the hearer does infer such a consequence, so arriving at an interpretation of the utterance. (Observe that an *abductive* process, in order to choose the [*doxastic*] *context*, is—as Dascal noticed in advance—the key piece of the whole procedure.)

III. SOME ISSUES REGARDING RESEARCH PROGRAMS ON LINGUISTIC PERFORMANCE

An obvious way to conduct an investigation on a given subject of study is to start considering its most general aspects, what may be called the *ontological category* it belongs to; and then to build stepwise, proceeding from more general to more particular features, a maximally specific picture of the phenomena that concern us (down to the differences distinguishing their observable varieties). I will try to sketch very rapidly how this route could be followed, point out how in each of its milestones careful attention should be paid to—empirical or otherwise—constraints enforced by available knowledge (or presumably "best theories") about it.

3.1 An Ontological Ladder

Linguistic performance is, trivially, a type of action. This obvious fact has been mainly emphasized, I think (if we ignore sociolinguistics proper), by *analysts of discourse*, and warrants studying the purposes, planning, execution, monitoring, etc., of such a performance. But this approach is unduly restrictive; we should consider it instead as an *activity*, a more or less temporally stretched way of functioning human beings that doesn't necessarily depend on conscious goals, i.e., that **may** proceed by bouts resulting, at least some of them, not from an interpersonal effect intended (be it illocutionary, perlocutionary, a mere turn-taking or a keeping the floor, etc.); it may well result, in many cases, from noticeably more modest cognitive processes, such as, *inter alia*, just a habit. (Think of talking to a tropical fish pet, of cursing while being alone, and perhaps even of some formulaic expressions—such as greetings—when more or less mechanically uttered.)

This simple move, that ranges linguistic performance in a continuum going from, say, digestion to carefully fixing up a much needed tool, may be objected to by saying that one needs only to consider the central, prototypical cases, since the other ones will be included automatically in the category (namely, *linguistic performance*) because of their similarity to the former. But accepting the prototypical character of fully conscious (and perhaps even communicative) linguistic action does not imply a commitment to disregard the general category where these and the peripheral cases belong; just the contrary claim is arguably the correct one. And in the case at hand such an *ontological* strategy allows for an appropriate place to conceptually locate the—to my mind, not at all implausible—possibility of *inner speech* having (at least sometimes) **no motives** at all, but only causes[26]—as it happens, *pace* Freud, with at least a considerable part of our dreams.

Now, in those cases where linguistic performance may be considered as a proper *action*, or series of actions, its being only a specific case of an *activity* predicts that the whole variety of purposes and goals that philosophers and psychologists have considered regarding actions (see, for instance, Brand [1984: Part IV], and Deci and Ryan [1985: Parts I-III]) may be in principle present in any single instance of such a performance, the whole aggregate having a complex (possibly multi-hierarchical) structure while in addition each of these aims or purposes may have any degree or nuance. (I think that this prediction is very obviously corroborated every day—especially at the psychoanalyst's coach.)

A second step would lead to a more refined category, but, as a consequence of it, not always applicable to a piece of linguistic performance; that of *communicative actions*. And to these apply (insofar as they are adequate) the pragmatic theories we reviewed above (see especially the discussion on "informative intention" and "communicative intention" in Sperber and Wilson [1986: §§6, 11 and 12], where the difference with the cases of non-actional linguistic activity I just suggested is most apparent).

I think, finally, that the vantage point adopted allows to solve in a natural way the dilemma that arouses when the intuitively unbounded variety of possible ways of using language (a Wittgensteinian intuition that the endless expositions by analysts of discourse seem to corroborate) is considered jointly with the also intuitively satisfying—but for details irrelevant here—very small number of categories we find in classifications of speech acts. For in a similar way to the step taken from *activities* to *actions*, a second step (pointing, though, in a different direction than the one mentioned in the preceding paragraph) may take us from the latter to *acts*—which, taking my lead from Bühler, I submit are types of actions culturally specified and identifiable.[27]

3.2 Some Constraints

Two strongly opposing but undoubtedly unavoidable constraints must be considered in any theory of linguistic performance. As hinted at when discussing Wilson and Sperber's theory, they are mandatory not because they necessitate hypothesizing a detailed mechanism able to satisfy both, but in order to initially formulate global theories that **may** be compatible with them.

On the one hand, research alluded to in §2.2.1(mainly due to W. Marslen-Wilson) has convincingly shown that the interpretation of sentences is an extremely fast phenomenon, that in some individuals may lag only a couple of syllabs behind physical perception. This surprising discovery is one of the main drives for the creation of grammars that confer structure to sentences, and foresee their semantic interpretation, in an "incremental" way, that is, beginning at the very start of the sentence and proceeding along its temporal course ("from left to right," as it is usually said) down to its end: so-called combinatory grammars (Crain and Steedman [1982], Ades and Steedman [1982]; for recent statements and applications, see Haddock [1987], and Altmann and Steedman [1988]). And notice that whether grammar itself attributes structure "from left to right" or it is only the parsing-interpreting mechanism which does it, the important issue is that the interpretation **must proceed** in this manner, and with the speed just pointed out.

On the other hand, as Sperber and Wilson take pains to show, it is unquestionably necessary, in order to arrive at a pragmatically appropriate interpretation, to effect some or other type of inference from a comparatively small number of appropriate assumptions (a [*doxastic*] *context*), which must have been selected (in many cases, even, apparently, constructed) with a view to this very interpretation. But we know that inference in general increases its computational complexity in an almost explosive way as the task is more complex, and the need to introduce new premises, i.e., to execute an *abduction*, makes it even more explosive (as pointed out by Hobbs, Stickel, Martin and Edwards [1988]). It is necessary for the theory, then, to assume some kind of structure in encyclopedic knowledge that allows in a simple way to find (or construct) premisses needed for the interpretation. The classical solution to this problem was assumed for some years to be *frames*, in the sense of Minsky (1974)—a means of avoiding the so-called *frame problem* in AI—and other similar resources, such as the *schemas* and *scripts* of Schank and Abelson; but these devices have shown not to have the flexibility that was expected of them (*pace* Thagard [1984]), so that researchers either eventually abandoned them or postulated both more and more types of schemas, and links between them of an ever increasing complexity (an examination of Schank [1981] is very instructive in this regard). A different tack is followed by *connectionist methods*, that avoid postulating specific data structures but achieve results (for some cases) identical to those obtained with structures such as schemas. Unfortunately, for the moment it seems that the claims that had been advanced regarding their capabilities were greatly exaggerated (see Pinker and Mehler [1988],and even Nadel *et al.* [1989]), so that other ways of solving the problem must be found;[28] what cannot be done, though, is to ignore it.

3.3 A Hypothesis of Commonality

If, as Dascal assumed, both higher types of linguistic performance, communicative and non-communicative, have a single essential nature, then it seems natural to hypothesize that it is not necessary for the encyclopedic knowledge to be modified from the outside, driven either by new perceptual data or by an attempt to reach an interpretation of some pre-existing sentence(s). If (at least momentarily) this is accepted, a second hypothesis seems appropriate: that that knowledge must have a rich structure, since only by assuming such an internal complexity, this distinction of internal "parts," can this body of assumptions or "encyclopedic knowledge" be modified independently of matters of abduction.

Now, two questions seem important in this regard. Is the body of encyclopedic knowledge in itself static (albeit possibly poised in an unstable balance because of its internally heterogeneous nature), and its changes are an effect of some other (cognitive or non-cognitive, e.g., emotional) mental modules, or is it, on the contrary, "naturally" changing (and appearances to the contrary would be deceptive)?[29] The second question that immediately arises is, What structure must be assumed, depending on the answer given to the previous question? In other words, what kinds of internal differences would be able to drive by themselves reorganizational changes, without—or with a minimum of—external impingement of new "data"?

Finally, it is mandatory to specify, at least in a general way, both how the assumed changes, both internally (if any) and externally driven, are effected and what conditions must be complied with by any "agent" of the modifications—so that the semantics of linguistic expressions, that are not the only "agent," but certainly (possibly a part of) one of them, would be, in some way or other, constrained by the theory.

3.4 A Methodological Reminder

As it is clear, on pain of empirical vacuity, no putatively empirical theory may be accepted unless some of the links in the explanatory mechanisms it suggests are in principle accessible to observation, however indirect. This is a constraint not only necessary to impose, but also apparently feasible, since the research methods developed in empirical psycholinguistics (shadowing, priming, gating and so on) seem to be able to discriminate between different hypotheses as to the context of assumptions used in a particular situation to reach a particular interpretation of a given utterance.

APPENDIX

A Critical Examination of the Theory of Relevance

1. Outline of the Theory

The theory's name is taken from the admonition constituting the third conversational principle of Grice's; for Sperber and Wilson believe that it is the only principle needed in a satisfactory account of linguistic communication —they criticize what they think is a superfluous positing of several principles by Grice, and deem even more unparsimonious the addition of new principles (e.g., of politeness, and even irony) suggested in more recent work, such as Leech (1982). In a word, they demand a *principled*, motivated introduction of principles of (linguistic or general) communication, as against an *ad hoc* proliferation of them.

The theory was developed in a series of papers (Wilson and Sperber [1980a], [1980b] and [1986], and Sperber and Wilson [1981a] and [1981b]), and found what may be considered its "classical" formulation in Sperber and Wilson (1986); a more recent essay (Wilson and Sperber [1988]) introduces somewhat subtle changes. I will base my presentation on the "classical" version, while incorporating the main changes noticeable in the newest version.[30]

1.1 Initial Issues

In real communication acts there is always a halo of vagueness, of imprecision; they may be interpreted in different ways, at different levels of "deepness." This fact is an indication that what the communicator intends is not to simply forward the right interpretation of such an act to the communicatee, but to modify (some parts or aspects of) his/her conception of reality—what linguists use to call *encyclopedic knowledge*, although in fact it is just the sum total of his/her conception of reality—what linguists used to call encyclopedic knowledge, although in fact it is just the sum total of his/her assumptions about reality. (In acknowledging this, Wilson and Sperber contend,[31] it becomes possible to account in a most natural way for the further fact that many—if not most— sentences are not to be taken in its literal sense, but in a loose, figurative or even metaphoric sense).

Since there is **no** way for emitter and receiver of communication acts to **know** mutually what Grice's meaning theory posited to account for communication, they can only **assume** to have a mutual *cognitive environment*, i.e., each would find some assumptions (more or less) *manifest* to him/her[32] and assume that they are also (more or less) *manifest* to the other. In addition, Sperber and Wilson point out that even what according to Grice *is said* must be interpreted, inferred—from linguistic evidence and the assumed cognitive environment. Finally, they seriously question both the obscure theoretical status of the ways (only shown by example) suggested by Grice as methods to inferentially arrive at implicatures, and—as mentioned above—the multiplicity of principles that Grice invoked in his account of such inferences. Finally, they claim that it is necessary to give an explicit theory of the inferences by means of which the receiver (as regards linguistic communication, the hearer) arrives at an interpretation, a theory that should be formal[33]—in fact, a logical theory.

1.2 Main Ideas

The final goal of human cognitive faculty is to improve as much as possible the individual's knowledge of his/her environment with a minimum processing cost, so that all human beings **automatically** "aim at the most efficient information processing possible."[34]

Now, when we face a cognitive task we have available two kinds of information, one already at our disposal (*encyclopedic knowledge*) and the other provided by the task itself—e.g., in a linguistic communicative action, by the sentence(s) uttered. Usually we take advantage of both types of information, and so we are able to obtain a cognitive effect (such as adding new beliefs, changing, deleting or modifying the degree of confidence in old assumptions, etc.), an effect that was not to be obtained from any one of those two types of information by itself. The **relevance** (for the individual who processes it) of the processing of a piece of information increases and decreases, respectively, according to the increase in cognitive effect achieved and the increase in effort needed to process it (in both cases, assuming a *ceteris paribus* clause).[35] So, we can say that humans aim *automatically* at a maximum of relevance, i.e., to secure as great cognitive effect as possible with a minimum processing cost.

Every communicator (a speaker, e.g.) conveys with his or her communication act a presumption that this act is to some extent relevant to the communicatee, (i.e., that its effects will be at least adequate to the latter and that the effort needed to obtain them will be minimum), since otherwise it would not be attended to. This fact can be expressed saying that every communicative act conveys a presumption of being consistent with the **Principle of Relevance**. So (as regards linguistic communication) the speaker will aim at wording his or her intervention in such a way that the **intended** interpretation is consistent with such a principle, while the hearer will aim at **obtaining** the—or at least an—interpretation consistent with it. (Both, then, will observe the principle, each in his/her own interest.)[36]

It would be utterly impossible, and also there would be no point in using the whole *encyclopedic knowledge* of the individual as (complex) premise of the inference leading to the interpretation of the communicative act. In every case, then, a part of it would be selected (and possibly completed with new assumptions) in order to infer such an interpretation. In other words, a much smaller and appropriate **context** (of assumptions) will be constructed, that combined with the new information provided by the "content" of the act itself will form the (complex) premise of the inference leading to the interpretation of the act.

1.3 Specific Aspects of Interpretative Inference

Although Sperber and Wilson acknowledge that our encyclopedic knowledge includes some "organizers," such as *frames*, *schemes* or *scripts*, they apparently assume that essentially it is a huge collection of *assumptions* of propositional format; i.e., they think about it as consisting essentially in assumptions whose content would be either fully specified, complete propositions, or parcially specified propositional forms ("incomplete propositions"). And so the rules of mental logic can apply directly to them.

We must assume, they think, that such a logic is simply a "natural deduction" formal logic,[37] but where trivial consequences are not derivable (because of the fact that nobody derives them in real life, when not making logic exercises); and they accordingly modify a popular system of *natural deduction logic* so that these consequences are avoided.

2. Problems and Difficulties

The approach adopted in relevance theory allows Sperber and Wilson to criticize effectively previous detailed attempts at explanation of linguistic communication phenomena[38] and to tackle successfully most if not all problematic cases discussed in the literature (plus a number of others they introduce themselves). Certainly, they do not present a full theory of the inferential processes they attempt to explain (as just mentioned, they only offer some modification of a previously existing logical system), but given the preliminary nature of their exploration of the theory it would be unreasonable to demand any more detailed developments. Nevertheless, I will advance some objections to it; some of them (which will be dubbed "internal") are to be considered as proper criticisms, since they assume the legitimacy of the field of study of this theory as an autonomous domain of investigation; while the other ones ("external" problems) look at the theory as a first step towards a satisfactory theoretical account of linguistic performance in general—an ultimate goal which, however, it is not at all obvious that D. Wilson and D. Sperber ever aimed it.

2.1 Internal Problems

In a preliminary way it may be pointed out that the assumption of the absolute advantage of having a single explanatory principle instead of a multiplicity of them is open to a number of critical observations. On one side, rarely, if ever, is there really one single principle able to explain every phenomenon in the domain of analysis (usually there are all kinds of auxiliary hypoteses—not only of a "bridging"nature—supplementing it),[39] so that claims to this effect should be critically scrutinized in order to check their correctness or otherwise (we will see below that under the appearance of not needing but the principle of relevance lurks an important conceptual equivocation). On the other side, the methodological example of generative transformational grammar, which Wilson and Sperber like to have recourse to, clearly point in the opposite direction since the famous controversy on the *goals of linguistic theory* (a methodological approach now fully instantiated in Government and

Binding theory); whenever the phenomena to be explained are really complex (and it doesn't seem to be much simplicity in human linguistic communication) a small number of different principles operating on a single field may account in an "economical" and non *ad hoc* way for a wide variety of empirical effects.[40] Finally, the phenomena that at least one of the offending principles recently introduced in pragmatics, the principle of politeness, tries to explain are not addressed at all by the originators of the theory of relevance,[41] so that one must suspend judgment on its adequacy until some attempt at their explanation is offered. Let us see now more substantive objections.

2.1.1 General Objections

First, it is not at all obvious that the fact that we would like to arrive at true conclusions when starting from true premises implies that the student of human mind, in order to model this ideal or aspired after *performance*, should assume underlying it, as *inferential competence*, a strict logic. Researchers in psychology of reasoning have found for more than fifty years that human beings' everyday inferences are notoriously unreliable even when errors are hard to attribute to extraneous factors such as memory gaps, processing difficulties and so on.[42] And it is out of sheer impossibility—or so it seemed to them—to account for such errors while maintaining the hypothesis of an underlying logical competence that some of these researchers eventually proposed a different (non logical) type of mental inferential faculty—a faculty or capacity that under very special conditions would result in making correct logical deductions and that would also account for logical errors incurred in less ideal circumstances. As to the argument that preservation of truth under inference was presumably a necessary condition for survival of human species, it may be countered noting first, that in fact there are suggested inferential systems that, although not being of logical nature (i.e., not formal), [43] are able to preserve truth (while also predicting observed types of logical errors), and second, that—as it has been repeatedly pointed out—[44] what really must have been necessary for survival is that, in the long run, a *maximum* of conclusions arrived at is appropriate (conducive to right action, etc.) in *real*, not abstract or formal, situations encountered. (A probabilistically optimum adjustment to ecological—including psychological—patterns of event development is undoubtedly more important for survival than unexceptional adjustment to logical deduction patterns.)

Another dubious feature of relevance theory, and directly connected with the previous one, is its conception of encyclopedic knowledge as an (essentially unstructured) aggregate of propositional forms—a feature certainly very convenient when it is assumed that the inferential resources available have the form of a (natural deduction) logic. I will not rehearse here the arguments pro and con propositional representations of knowledge (even though I believe that certain empirical findings make almost mandatory a rejection of the **general** validity of such an assumption); nevertheless, by unquestioningly assuming that such is the form that our encyclopedic knowledge *always* has, Wilson and Sperber not only are open to serious objections, but strain our most basic intuitions about certain kinds of knowledge such as musical skills, knowledge of human faces, of instanciated artistic qualities, etc. (that undeniably may play a critical role in the interpretation of sentences, and so should be able to be part of the input to the mental logic that these writers postulate).[45]

It is possible that one of the sources of Sperber and Wilson's assumption of both a propositional encyclopedic knowledge and a logical system of inference is their virtual confinement to strictly informational communication, i.e., to dialogues (when they deal with linguistic communication where the sole finality of the interchange is to provide some (more or less clearly demanded) information.[46] In order for the theory of relevance to be possibly considered—as they manifestly aim to—a satisfactory general theory of linguistic (and other types of human) communication, they should tackle dialogues of a different type (which, on the other hand, have no reason to be less literal or "serious" than those they study): narratives, idle conversations (such as those unendingly examined by the *founding fathers* of discourse analysis), some ritualized types of intercourse,[47] "showing off" use of language (so effectively emphasized by Goffman), speaking for the sake of it, etc.[48] And their examples (they follow the pragmatic tradition of arguing for a theory of speech interpretation by showing how it may account for specific examples of extremely brief—most times just one short sentence long—linguistic interventions) should include more extended examples, where, as analysts of discourse have convincingly shown, meaning *negociation* may enter essentially. Until they so extend the scope of their theory I don't think it may be considered a candidate to the high explanatory position they claim for it.

But all these difficulties are comparatively mild when the theory of relevance is confronted with what I think are its harder internal problems, i.e., the very concept of *relevance* propounded and the way of reaching a sentence's interpretation that it is suggested.

2.1.2 Central Objections

As we saw, Sperber and Wilson advance a *comparative* concept of relevance differing in important ways from the usual one; (a) it is not relational, and (b) it is (I would say) quasi-quantitative. As to the first difference; it is true that in their "classical," 1986 presentation they extensively demonstrate, through a reanalysis of discussed examples and the study of new ones, that the Principle of Relevance can without loss substitute for all four principles[49] postulated by Grice. But this putative demonstration relies both on the common notion of *relevance* and on their technical, comparative notion, in such a way that most of the arguments equivocate on this double use—their appearance of being convincing comes from a continuous shift between both ways of understanding the word "relevance." I do not have space here to show example by example how this equivocation lies under their whole "overcoming" of Gricean principles, so that the reader must check it by himself/herself. (I will discuss immediately the newer, 1988 presentation.)

The significant issue, though, is that, regarding the "classical" presentation, if the relationship alluded to by the ordinary notion of relevance is not included (in some way or other) in the theory, it is impossible, in general, to uniquely determine the interpretation of an utterance. For in their analysis of selected examples of linguistic intercourse, Wilson and Sperber either use expressions such as "the plausibility of some [interpretative] hypotheses may already be manifest in the environment" (Sperber and Wilson [1986], ch. 3, §8, examples (64), discussion immediately before it),[50] simply assert that the interpretation is unique, or "explain" the offered interpretation by presenting a *context* (of assumptions) supposedly constructed by the hearer but that in fact is so chosen that it leads uniquely to that interpretation. Now, the interpretations they choose for their examples seem eminently plausible, and it is very probable that they are those that would occur to anyone in the situations they describe. But the reason for this fact is that **we**, as participants in the culture their characters belong to, **know** what their most probable intentions, desires or tendencies to act, think, etc., are, and also choose interpretations that are *relevant* (in its normal, non-technical sense) in relation with them; while what Sperber and Wilson apparently **believe** they are showing is that these interpretations are chosen because they **are optimally relevant,** in their technical sense of this expression—i.e., because they are the ones that include a maximum of conclusions and are arrived at with a minimum of effort. (As to the context of assumptions and how it is, in its turn, constructed, see below.)

What happens with the new, 1988 version of the theory of relevance? Now it is suggested that the interpretations first occur to the hearer because of his or her knowledge of—or better, assumptions about—the situation (presumably including the needs, habits, interests, etc., of the speaker), and that such interpretations are, as if it sere, checked afterwards for their conformity with the Principle of Relevance, i.e., for *minimum* effort and cognitive *adequacy*. This means that the usual *concept* of *relevance*, with its relational character, is here simply **shifted** (and not against its ordinary sense, because of its very vagueness) to the expression "cognitive adequacy"; as to the *Principle* of *Relevance*, no matter its validity or lack of it,[51] does **not** operate as an **explication** of the concept of relevance: its name is itself arbitrary, irrelevant, since its whole content is given in its (new) definition—the one I recorded above— where the word "relevance" could be replaced by any of the host of others ("importance," "effect," "saliency" and so on) or by any neologism, with no change in import nor significance for the theory.

It may be thought that these are mere terminological quibbles. But in fact it is not so. For we have just seen that if one holds to the "classical" version, Sperber and Wilson's *explanation* of the interpretations reached by the hearer amounts to a mere assertion that they are just the ones optimally relevant in the situation, and to a hidden appeal to the intuition of the reader on matters of relevance (in its ordinary, unexplicated sense), not to an explanation of it (but for the idea of deduction from a constructed *context* of assumptions, to which I will return in the next paragraph). And if the more recent version of the theory is adopted, either the situation is even worse, for the burden of explanation, after giving up the notion of mere quantity of consequences, is simply passed over to the nebulous term "cognitive adequacy" (that pending a detailed characterization cannot explain anything), or we still cling to the explanatory device of deduction from a *context*, with its attendant difficulties—that I examine directly below.

Let me comment briefly on the notion of *construction* (by the hearer and, in an anticipatory way, also by the speaker) *of a context* of assumptions *from which to infer* deductively *an interpretation*, either *relevant* or *cognitively adequate* (depending on the version of the theory). Now, the problem with this concept lies in the fact that, no matter which specification is chosen (either "relevant" or "cognitivelyadequate"), it must have some content, it must constrain in some way or other the eligible interpretations. This means that such a context must be selected from the total contents of the encyclopedic knowledge of the hearer (or, as regards the speaker's performance, from the one attributed to him/her by the speaker) and possibly augmented with some conplementary assumptions, so that relevant/cognitively adequate interpretations are inferable from it. But how is this constructed construction effected? Nothing is offered by Wilson and Sperber on this count; in fact, the problems attendant to that construction (an *abduction*, by the usual terminology)[52] are not even mentioned. It is true that they answered early criticisms (Gazdar and Good [1981]) about computational difficulties liable to arise in any implementation of their model of linguistic interpretation; but in fact (see Sperber and Wilson [1981c]), on the one hand, they failed to notice the problems of *selection* of assumptions from the encyclopedic knowledge to form what they call the *context* and, on the other, they dismissed all computational problems likely to arise on the grounds of their not attempting to present an empirical model of the on-line computational mechanisms involved in such an interpretation. Now, this line of reasoning flatly ignores the obvious point that a theory about general aspects of a real (not an abstract) system's functioning **must not** be such that the processes it hypothesize are unfeasible; and that if it does, it is unacceptable.

In a word, it may be said that, **but for the condition on minimum effort** (see, though, footnote 40 above), relevance theory is, in its present form, a *no theory* of linguistic communication, in that **it relies entirely on its readers unexplicated intuitions for its appearance of explaining such type of communication.**

If this diagnosis is correct, how was it possible that a piece of research so thoughtfully and carefully done, so admirable in many aspects and that by now has shown to be so fruitful[53] be fraught with such problems? I think that the reason why these went unnoticed is of a methodological nature. In keeping with usual practice in pragmatical investigations about non-conclusive inferences (implicatures, "invited inferences," etc.), Wilson and Sperber accounted for interpretation of offered examples by positing both the premisses from which the interpretation should be concluded (the so-called *context* [of assumptions]) and this interpretation itself; apparently, no empirical check nor test seems to be called for. But by operating in this way they merely achieve compatibility with pre-theoretical intuitions, i.e., no more than intuitive plausibility (in fact, if my argument is correct, to say that they achieved such a compatibility is a truism, since their account was essentially no more than a reformulation of these intuitions). Now, other theories may also result in plausible accounts, as Grice's original theory of conversation manifestly shows, no matter the limitations that in some cases (as regards explanation of irony, for instance) may be found in it. So, without a way of empirically testing them, theories of linguistic communication may suggest arbitrary explanatory devices and ultimately are not empirical theories, but formal exercises.

2.2 External Issues

Let us assume that relevance theory were a satisfactory account of (linguistic) communication—albeit in its first phases of development.

Clearly, this assumed fact would not be equivalent to its being also a general account (either satisfactory or not) of linguistic performance, since some "uses" of language are not communicative, at least in the ordinary sense of this word. (It is immaterial what position is adopted regarding the once heated debate on the identity or otherwise of "higher" thinking—such as reasoning—and subvocal speech, for monologues patently exist, and even some types of linguistic action performed before—and apparently directed to—other persons are not proper communication.)[54] Now, since nowhere in their presentations of the theory of relevance do Wilson and Sperber try to account for linguistic performance in general (as mentioned above, immediately before §2.1), the following comments should not be considered properly as criticisms, but as observations on the problems that would be encountered by any attempt at extending such a theory to cover the whole of linguistic performance.

Specifically, it seems very doubtful that the suggested principle of (optimum, or maximum) relevance should be deemed valid for monologue and thinking—unless, i.e., one accepts in its most literal interpretation the time-honoured dictum that thinking is a dialogue of the mind with itself. Such an interpretation should include the assumption of this mind or soul neatly cleaving in two separate parts that would play, in an alternative way, the roles of speaker and hearer—the former part anticipating what information would be acceptably interesting, worth of attention on the side of the other part, and the latter inferring a maximally relevant interpretation of the expression "heard"—a double sided activity that does not seem especially attractive. (I do not mean that, were the principle of relevance adequate, it would be impossible to modify it so that it would be also applicable to the cases at hand without the need of this strained interpretation. My only point is that such a change **had** to be made in order to arrive at an acceptable theory of linguistic performance in general.)

Finally, it should be noticed that non-communicative linguistic performance may be, as a matter of fact, so different from communicative performance that, *pace* Dascal, two separate theories should be developed for them (e.g., with different principles of relevance); although it seems probable—but it is barely possible that this intuition be deceptive—that such pair of theories would show, if adequate, a high degree of parallelism. However, from a methodological standpoint it is in principle preferable to build a single theoretical frame for both types of phenomena (that apparently are very closely linked), introducing differences only if and where they are necessitated by the facts to be explained. This methodological advantage sufficiently motivates, I think, a strategy of attempting first to develop a unified and adequate theory of linguistic performance, no matter its type; I suggested in the body of the paper, section 3, some initial issues that it seems important to examine in research towards such a theory.

NOTES

*. It is a pleasure to acknowledge the immense encouragement and help received from Mª Teresa Espinal, who for many years criticized my often grossly misdirected work on semantic and pragmatic topics and made generously her best to remedy my all to o frequent misinformation on the relevant literature. I am in a particular debt with her for criticism of the present paper; a criticism, alas, that I have only minimally paid heed to, since otherwise I would have to write quite a different (and, no doubt, much better) paper - a not small task that I prefer to defer to a different occasion.

To the gentle sharpness of Mª Dolores Avia, who upon reading a previous version immediately spotted its weakest—and, of course, most convoluted—section, and graciously managed to raise in me an interest to read it in a dispassionate way, I owe the privilege of having been given the opportunity to discover its misleading arguments *all by myself*, my readers—if any—the relief of having been spared a confrontation with it in that horrid version.

1. I think that memories of the not so distant era of behaviourism, when psychology, sociology, linguistics and philosophy were in a happy agreement on main issues (as to philosophy, there seems to be still lingering much of the old ban to matters where fact is immediately apparent), could be somewhat sobering.

2. I am perfectly aware that such a presentation of previous ideas and research will look suspiciously like an instance of the so much and so justly decried (by historians) *backward history*. But I have no intention at all of sketching any history, however summary, of the studies in my field of interest; I will just attempt to highlight a few important steps whose knowledge seems to be necessary to properly situate my own work and aim it in the right direction. (Once again, a replay of *Historia magistra vitae*.)

3. This term should be taken in a generalized sense, since it is meant to include even those "parts" of the mind that by definition lie outside of any module in its conventional sense (i.e., any distinguishable subtype of operations within the "central systems"—notice the plural form used by Fodor himself—that would operate on the outputs

of the several encapsulated *modules*). On the other hand, all talk of "modules" in this paper is for convenience; I take no stand here on the present controversy about their functional size, interconnection, degree of encapsulation and so on.

4. I will drop from now on this qualification, that should be construed always to apply.

5. And other factors, undoubtedly, such as the reversal carried through by Chomsky on the prevailing primacy of production—a primacy that effectively almost excluded the other type of performance from the minds of linguists when they dealt (almost always in passing, anyway) with linguistic activity. This reversal was rapidly stamped on psycholinguistics based on early generative grammar or taking its lead from it, which is the source of almost every psycholinguistic investigation from around 1970 to this day.

6. McNeill (1977), chs. 4-8, provided an analysis of the conceptual structures he took to underlie the linguistic structures of utterances and form, in the productive linguistic performance process, their starting point (he has since abandoned this suggestion, claiming instead that *inner speech*, which would be such an starting point, includes—in addition to a *global-sybthetic, imagistic* form of thinking—a strictly *linguistic* form of thinking as syntactically structured as *outer language* and so irreducible to something else; see McNeill [1987]). Leaving aside the interesting fact that these old analyses were essentially identical to the conceptual analyses of sentences independently advanced, on linguistic grounds such as the notion of lexical decomposition, by Jackendoff (1983), it should be noticed that even when he suggested them, McNeill acknowledged that they were essentially unsatisfactory due to their not being in any way connected to the goals and purposes of the speaker.

7. It is true that, as pointed out in Winograd (1987), such resources and operation work in an entirely *ad hoc* way, so that no illumination results; but this observation is of small moment for our present purpose, as we will see later, when criticism of *relevance theory* will be offered. (For a more "run-of-mill" example of computer models of language production, cp., Arbib, Conklin and Hill [1985], part IV.)

8. Concerning Searle, he only rather lately, after his pragmatic views were formed and fully developed in the "analytic" entourage, seems to have run into Husserlian ways. A check at the dates of the papers collected in Searle (1979), before the turn that emerged in his (1983), is revealing on this matter.

9. In all fairness it should be noted that recently a few of them seem to be feeling some urge towards a more theoretically bent type of research. See, e.g.—to quote from a single source—Garrod and Sanford (1987), Ariel (1987), and especially Vicher and Sankoff (1988).

10. A few scholars think—or at least thought—that semantic and pragmatic constraints allow dispensing with such syntactic task. But Frege's (and, later, Chomsky's) arguments for compositional semantics are, no matter how many minor corrections and conditions of application one adds to them, apparently impossible to reject wholesale, so that a consensus seems to have been reached about the *general* need to carry out such a *parsing*.

11. I disregard adumbrations such as those of Fodor and Garrett (1966)—also voiced in the attendant discussion.

12. Notice that the observation made immediately before §2.2.1applies directly to theories described in this section.

13. Which, of course, are refined in a series of papers by Grice himself (until his [1982]) and other writers.

14. He pointed out later how the idea was immediately taken up by Dummett and other young Oxford philosophers. (See Dummett, [1974] and [1976], where this idea becomes definitional of semantics; cp., also most of the other essays of Evans and McDowell [1976].)

15. A notion that is certainly lurking in Davidson (1966); but notice its overt formulation in a lecture delivered in the same course of lectures than his (1974), i.e., Anscombe (1974).

16. The need for a theory of "semantic structure [since] the interpretation of utterances of complex sentences will systematically depend on the interpretation of utterances of simpler sentences, for example," is acknowledged for a full theory of interpretation —in my sense—since Davidson (1973b: 16th paragraph from the beginning). In other words, he acknowledges that for that case some kind of parsing is necessary (to comply with Frege's compositional principle, no matter how broadly construed), although he just assumes it is somehow carried out.

17. See the references given in Clark and Clark (1977), chs. 2 and 5.

18. An early review of work in the general (psychologists') approach here dealt with may be found in Just and Carpenter (1977); more recent works are, e.g., Flores d'Arcais and Jarvella (1982) and Kurcz, Shugar and Danks (1985). For a characteristic example of computer theorists' approaches to language understanding see Arbib, Conklin and Hill (1985), ch. 8.

19. Maybe it would be apposite to notice that they are a philosopher and a linguist. A not too dissimilar combination will be found as regards research we will review in a moment (i.e., Sperber and Wilson's theory of relevance), due to an anthropologist and a linguist.

20. They separate out so-called conventional illocutions (which other writers dubbed "arch-performatives"), i.e., those that depend on the an appropriate social convention to effectively change (social) reality—those through which people "do things with words" in the strongest possible sense, such as christenings, proclamations, verdicts etc., According to them, in carrying out these linguistic acts, certain conventions take the place of the "reflexive intentions" (causing an effect on the hearer by means of the recognition of just such an intention: Grice [1957]) that operate, based on a communicative presumption, in the other cases of linguistic communication.

21. Notice that they reject the notion of *mutual knowledge*, much discussed and variously named, but in some form or other always assumed in the tradition coming from Grice (1957). And they do so because, they argue, there is no way to get at a *knowledge* in these matters —there can only be presumptions, beliefs.

As Fox and Geis (1984) point out, one of the first pieces of research in linguistics where an inferential task from assumptions is suggested as a generally necessary means of arriving at a correct interpretation of a sentence is Lakoff (1969). But it should be kept in mind *a)* that such a proposal concerned the **grammaticality** of a sentence (that was viewed, then, as **relative** to these assumptions), not just its semantic interpretation, and *b)* that these were assumptions of a specific, limited type -those "carried" (or 'invited') by the sentence itself, i.e., what since its rediscovery by Strawson was known by the name of *presuppositions*.

22. First of all they suggest a *linguistic* presumption, that includes *a)* (of course) knowledge of the language, *b)* that the fact of having perceived the sentence uttered would allow the hearer (one may presume, by the use of lexical and syntactic -structural-information) to arrive at the reading or readings of the sentence (that generally -as indexical show- will be more or less undeterminate), and *c)* the feasibility of identifying what the speaker is saying by making use, in addition, of mutual contextual beliefs. Then they advance a presumption of *communicative* intention by the speaker, a *compatibility* presumption condition between the form of words used and the illocuationary act attributed, by virtue of its utterance, to the speaker (i.e., the specific communicative intention assumed), and a cancelable presumption of *literalness*.

23. Unfortunately, while familiar with relevant research by many linguists, he apparently is not aware of Bach and Harnish's work.

24. No matter this similarity, they notoriously ignore the latter's theoretical advances—mainly, it seems, because of not having entirely discarded the notion of *mutual knowledge* and having kept to a plurality of principles of conversation.

25. And that perhaps would be rejected by these writers if explicitly formulated, even though all presentations of their theory are compliant with it.

26. I eschew the philosophical controversy on reasons and causes, that here would be otiose.

27. Notice that Austin used interchangeably "illocutionary action" and "illocutionary act"—as it should be since he paid no attention to cross-cultural comparisons nor differences. Notice also that a way of culturally making a distinction between types of actions is to tie it to a distinction between procedures or norms in an institution; and since "the tongues of men" are institutions (of course, culturally fixed), different entries in the lexicon of a particular language for these actions do, in principle, effect the finite cleavage in *acts* of the unbounded variety of illocutionary actions just alluded to. On the other hand, these *acts* may be later grouped together in a few broad categories (an operation that Austin executed, as everyone knows, on the base just of such entries).

28. An interesting attempt to find a "middle way" in the sense just hinted at may be found in Wilensky (1986).

29. Of course, if the latter is assumed, some account must be given of the deception.

30. Since the 1988 paper is, understandably, much shorter than the 1986 book, it is not easy to tell whether certain parts of the doctrine not mentioned in the former are to be considered as eliminated or at least downplayed by its authors, or as maintained with their previous role but are simply omitted in order to concentrate on more basic or controversial issues, or on some aspects in need of qualification.

31. See Wilson and Sperber (1988), §4.

32. "Manifest" means here, for Sperber and Wilson, easily accessible (to the mind), e.g., because of perceptual evidence.

33. They point out the enormous success of generative linguistics, an approach based on a thorough formalization of its field of study.

34. Sperber and Wilson (1986), ch. 1, §9, p. 49.

35. I merged the definitions found in *ibid.*, p. 48, and in Wilson and Sperber 1988, §3, p. 140 (keep in mind what I said in a previous note about differences between these two presentations of the theory). Notice, on the other hand, that this definition of *relevance*, a quasi-quantitative one (Sperber and Wilson call it "comparative") does **not** agree with pre-theoretical intuitions—or so it seems to me. In its usual sense "relevance" is always relative to a matter, an issue, I would even say a purpose, something aimed at; and points to the feature of potentially furthering or being conducive to reaching it.

36. Here there seems to be a clear revision of the ideas expounded in the 1986 book (cp., Sperber and Wilson [1986: ch. 3, §7, (62), p. 158] and Wilson and Sperber [1988: §3, p. 140]); accordingly, the second version is virtually the only one presented here.

37. Sperber and Wilson argue that (a) speakers and hearers obviously want to arrive at true conclusions when starting from true premises, and this is exactly what formal logic aims at being a theory of, and (b) the fact that logic is a highly developed discipline should encourage us to postulate a use of its methods when constructing a theory of the (inferential) mind. As to the *natural deduction* nature, it seems the obvious choice once accepted the previous points.

38. The qualification "detailed" is aimed at excluding other, more general or "philosophical" studies—such as Dascal (1983), which, as I previously mentioned, they seem not to be familiar with.

39. Arguably the whole of the *structural theory* of science is predicated on this observation.

40. Sadock (1986) raises extremely similar, if not identical doubts.

41. I take it that the mass of observations and analyses presented in the literature of linguistic "reflections" of politeness (see, e.g., Brown and Levinson [1986], first published as a part of Goody (1978), which is the version used by Leech [1982]) is enough to render unacceptable summary rejections such as the one offered by Sperber and Wilson. (For recent papers on these issues, see Holmes [1988] and Matsumoto [1988].)

42. Just one example: the universal tendency to interpret sentences of the form "If *A* then *B*" was warranting the conclusion "If *no-A* then *no-B*." It should be noticed that even professional logicians make these "errors" when the logical structure of the inference is carefully hidden under an everyday disguise.

Compare again a practically identical argument to the same effect independently advanced by Sadock (1986).

43. See Johnson-Laird (1983) and especially (1986); in fact, Sperber and Wilson mention this work, but they dismiss it cursorily as if its creation were a capricious quirk on this psychologist's part. A comparatively recent empirical study is Fox and Geis (1984).

Although they cannot be properly called "non logical," there is a host of attempts in AI and attendant cognitive psychology, since the very inception of so-called procedural semantics (Davies and Isard [1972]), to construct adequate inferential systems that, instead of having a logical foundation, are (computationally feasible and) able to account for our ability to (sometimes) infer logically. "Non-monotonic logic" may be considered as one of the latest celebrated attempts in this direction.

44. For a more extended argument see, e.g., Cherniak (1983).

45. Note that I am **not** claiming that these subjects, particularly artistic qualities and feelings, are ineffable—I remain uncommital to the well-known thesis of their ineffability.

46. I take it that the extension of their theory to just interrogatives and imperatives (but not to other non-declaratives), that they announce in Wilson and Sperber (1988: §6), confirms the informativeness bias of their approach. (Which in turn, I suspect, is a consequence of the very special nature, from the informational standpoint, of the profession of its authors -the academical profession.)

47. Such as the *ritual insults* studied by Labov and their equivalents in other (sub)societies.

48. Von Savigny (1988), §30, also notes that on many linguistic exchanges (where grammatically "declarative" forms are used) there is really **no informative intention**.

49. Five if you include the overarching principle of cooperation, which, on the other hand, it seems to me obvious that may be considered as an explicit form of the commitment by the members of a society to the principle of relevance. A commitment—comparable to Lewis' triangle principle—that is not otiose to posit (like the elephant supporting the elephant...), since in certain societies its validity is suspended, at least for some linguistic interchanges—see [Ochs-]Keenan (1975).

50. Even in cases where they are intent on showing clearly the vagueness of possible interpretations one can find accounts such as this one; "There is no precise assumption, apart from the one explicitly expressed, which [the speaker] can be said to intend [the hearer] to share. Yet there is more to [the former's] utterance than its explicit content: [the speaker] manifestly intends [the hearer] to draw some conclusions from what [was] said, and not just any conclusions": Sperber and Wilson (1986), ch. 1, §11, discussion of ex. (51). See my comments immediately following in the text.

51. I think that in its literal tenor (in the 1988 version, of course) it is too strong. For the enforced *minimum processing effort* condition would exclude the indefinite chain of new interpretations that in many cases come successively to mind when the hearer is really interested in "probing" what the speaker meant—when (s)he is prepared to spend more cognitive effort than a bare minimum. And I have no doubt that Wilson and Sperber would shun such an exclusion, since they clearly hinted at these various possible interpretations in the very first chapter (§2) of their 1986 book. On the other hand, the theory needs to be developed here, on pain of vacuity, since otherwise there is no way to know what is a *minimum effort*—the only determinate value would be just *naught*, which is nonsense.

52. As indicated in the body of the paper, §3.2, third paragraph. Add to what is argued there the observation that abduction, when having only available (to choose from) an essentially unstructured huge set of assumptions, confronts—as pointed out, e.g., in Cherniak (1983: §§8 and 10)—a problem similar to the problem of deduction first formulated in Lewis Carroll's famous parable of Achilles and the turtle (in fact, a computing intractability version of it). For in order to select the right premisses you must

first deduce from them conclusions that potentially comply with the desired specifications; but this was the very task you initially intended to execute, and the vast size of the set effectively prevents trying successively, "by brute force," all plausible premisses. (For a treatment of abduction—in a severely limited and highly structured setting—as a method of interpretation of sentences, see Hobbs, Stickel, Martin and Edwards [1988].)

53. See, for instance, Kempson (1986) and (1988a), Blakemore (1986) and (1988), Carston (1988), and even Wilson and Sperber (1988: §§4-5).

54. Malinowski's classical *phatic communion* behaviour, and Piaget's *egotistic speech* in children (no matter Vigotsky's criticisms, that do not affect the aspects of it we are interested in) come readily to mind.

REFERENCES

Ades, A.E., and M.J. Steedman, 1982: "On the Order of Words," *Linguistics and Philosophy*, 4, 4 (1982), pp. 517-58.

Altmann, G., and M.J. Steedman, 1988: "Interaction with context during human sentence processing," *Cognition*, 30, 3 (December 1988), pp. 191-238.

Anscombe, G.E.M., 1974: "The First Person," in *Guttenplan* (1974), ch. 3.

Appelt, D.E., 1984: *Planning English Sentences*; Cambridge: C. Univ. Press, 1985.

Arbib, M.A., E.J. Conklin, and J. Hill, 1985: *From Schema Theory to Language*; Oxford: O. Univ. Press, 1987.

Ariel, M., 1987: "Retrieving propositions from context: Why and how," *Journal of Pragmatics*, 12, 5/6 (Dec. 1988), pp. 567-600.

Bach, K., and R.M. Harnish: *Linguistic Communication And Speech Acts*; Cambridge, Mass.: MIT Press, 1979.

Berwick, R.C. and A.S. Weinberg, 1984: *The Grammatical Basis of Linguistic Performance: Language Use and Acquisition*; Cambridge, Mass.: MIT Press.

Bever, T.G., 1968-70: "The Cognitive Basis for Linguistic Structures," in J.R. Hayes (ed.), *Cognition and the Development of Language*; New York: Wiley, 1970.

Blakemore, D., 1986: *Semantic Constraints on Relevance*; Oxford: Blackwell, 1987.

Blakemore, D., 1988: "'So' as a constraint on relevance," in Kempson (1986b), ch. 8.

Bock, J.K., 1981: "Toward a Cognitive Psychology of Syntax: Information Processing Contributions to Sentence Formulation," *Psychological Review*, 89, 1 (Jan. 1982), pp. 1-47.

Bransford, J.D., and N.S. McCarrell 1972: "A Sketch of a Cognitive Approach to Comprehension: Some Thoughts About Understanding What it Means to Comprehend," in W. B. Weimer and D.S. Palermo (eds.), *Cognition and the Symbolic Processes*; Hillsdale: Erlbaum, 1974.

Bresnan, J., 1975-76: "A Realistic Transformational Grammar," in M. Halle, J. Bresnan and G.A. Miller (eds.), *Linguistic Reality and Psychological Theory*; Cambridge, Mass.: MIT Press, 1978, ch. 1.

Brown, P., and S.C. Levinson, 1986: *Politeness: Some universals in linguistic usage*; Cambridge: C. Univ. Press, 1987.

Butterworth, B. (ed.), 1979: *Language Production*, vol. 1: *Speech and Talk*; New York: Academic Press, 1980.

Butterworth, B. (ed.), 1982: *Language Production*, vol. 2: *Development, Writing and Other Language Processes*; New York: Academic Press, 1983.

Carston, R., 1988: "Implicature, explicature, and truth-theoretic semantics," in Kempson (1988b), ch. 7.

Cherniak, C., 1983: "The 'Universal Acceptance of Logic'"; Berkeley: Inst. of Cogn. Studies, Univ. of California.

Clark, H.H., and E.V. Clark, 1977: *Psychology and Language: An introduction to psycholinguistics*; New York: Harcourt Brace.

Cole, P., and J.L. Morgan (eds.), 1975: *Syntax and Semantics*, vol. 3: *Speech Acts*; New York: Academic Press.

Cooper, W.E., and J. Paccia-Cooper, 1980: *Syntax and Speech*; Cambridge, Mass.: Harvard Univ. Press.

Cooper, W.E. and E.C.T. Walker (eds.), 1978: *Sentence Processing: Psycholinguistic Studies Presented to Merrill Garrett*; Hillsdale: Erlbaum, 1979.

Cosmides, L., 1988: "The logic of social exchange: Has natural selection shaped how humans reason? Studies with the Wason selection task," *Cognition*, 31, 3 (Apr. 1989), pp. 187-276.

Crain, S. and M. Steedman, 1982: "On not being led up the garden path: the use of context by the psychological syntax processor," in D.R. Dowty, L. Karttunen and A.M. Zwicky (eds.), *Natural language parsing: Psychological, computational, and theoretical perspectives*; Cambridge: C. Univ. Press, 1985, ch. 10.

Cutler, A., (ed.) 1982: *Slips of the Tongue and Language Productions*; Amsterdam: Mouton.

Dascal, M., 1983: *Pragmatics and the Philosophy of Mind, 1: Thought in language*; Amsterdam: Benjamins.

Davidson, D., 1966: "The Logical form of Action Sentences," in N. Rescher (ed.), *The Logic of Decision and Action*; Pittsburgh: Univ. of P. Press, 1967, ch. III; reprinted in (1980), essay 11.

Davidson, D., 1971–74: "Psychology as Philosophy," in S.C. Brown (ed.), *Philosophy of Psychology*; London: Macmillan, 1974; reprinted in (1980), essay 12.

Davidson, D., 1973a: "Belief and the Basis of Meaning," *Synthese*, 27 (1974), pp. 309-23; reprinted in (1984), essay 10.

Davidson, D., 1973b: "Radical interpretation," *Dialectica*, 27 (1973), pp. 313-28; reprinted in (1984), essay 9.

Davidson, D., 1974: "Thought and Talk," in Guttenplan (1975), ch. 1; reprinted in (1984), essay 11.

Davidson, D., 1980: *Essays on Actions and Events*; Oxford: Clarendon.

Davidson, D., 1984: *Inquiries into Truth and Interpretation*; Oxford: Clarendon.

Davies, D.J.M. and S.D. Isard, 1972: "Utterances as programs," in B. Meltzer and D. Michie (eds.), *Machine Intelligence* 7; Edinburgh: E. Univ. Press, ch. 17.

Dean Fodor, J., and L. Frazier, 1980: "Is the human sentence parsing mechanism an ATN?" *Cognition* 8, 4 (Oxt. 1984), pp. 417-59.

Dummett, M.A.E., 1974: "What is a Theory of Meaning?" in Guttenplan (1974), ch. 6.

Dummett, M.A.E., 1976: "What is a Theory of Meaning? (II)," in Evans and McDowell (1976), ch. IV.

Evans, G., and J. McDowell (eds.), 1976: *Truth and Meaning: Essays in Semantics*; Oxford: Clarendon.

Flores d'Arcais, G.B. and R.J. Jarvella (eds.), 1982: *The Process of Language Understanding*; New York: Wiley, 1983.

Fodor, J. and M. Garrett, 1966: "Some reflections on competence and performance," in J. Lyons and R.J. Wales (eds.), *Psycholinguistics Papers*; Edinburgh: E. Univ. Press, pp. 135-54. (Discussion of the paper: pp. 154-79.)

Fox, R.A. and M.L. Geis, 1984: "Mental Logic and Sentential Comprehension," in J. Drogo, V. Mishra and D. Testen (eds.), *Papers from the Twentieth Regional Meeting, Chicago Linguistic Society*; Chicago: Ch. Ling. Soc., pp. 87-103.

Frazier, L. and J. Dean Fodor, 1978: "The sausage machine: a new two-stage parsing model," *Cognition*, 6 (1978), pp. 291-325.

Fromkin, V.A. (ed.), 1973: *Speech Errors as Linguistic Evidence*; The Hague: Mouton.

Fromkin, V.A. (ed.), 1977-80: *Errors in Linguistic Performance: Slips of the Tongue, Ear, Pen, and Hand*; New York: Academic Press, 1980.

Garrett, M.F., 1981: "A Perspective on Research in Language Production," in Mahler, Walker and Garrett (1981), ch. 10.

Garrod, S. and D. Sanford, 1987: "Thematic subjecthood and cognitive constraints on discourse structure," *Journal of Pragmatics*, 12, 5/6 (Dec. 1988), pp. 514-45.

Gazdar, G. and D. Good, 1981: "On a Notion of Relevance," in Smith (1981), ch. 2, "Comments and Replies," pp. 88-100.

Goody, E. (ed.), 1978: *Questions and politeness: Strategies ib social interaction*; Cambridge: C. Univ. Press.

Grice, H.P., 1957: "Meaning," *The Philosophical Review*, 66, 3 (1957), pp. 377-88.

Grice, H.P., 1967: "Logic and Conversation," William James Lectures delivered at Harvard University: partially published as "Logicand Conversation" [in Cole and Morgan (1975), pp. 41-58] and as "Further Notes on Logic and Conversation" [in P. Cole (ed.), *Syntax and Semantics*, vol. 9: *Pragmatics*; New York: Academic Press, 1978, pp. 113-27].

Grice, H.P., 1982: "Meaning revisited," in N. Smith (ed.), *Mutual Knowledge*; New York: Academic Press, pp. 223-43.

Guttenplan, S. (ed.), 1974: *Mind and Language*; Oxford: Clarendon, 1975.

Haddock, N.J., 1987: "Incremental Interpretation and Combinatory Categorial Grammar," *Proceed. of the Tenth Int. Joint Conference on Artificial Intelligence*; Milan, pp. 661-3.

Hobbs, J.R., M. Stickel, P. Martin, and D. Edwards, 1988: "Interpretation as Abduction," *Proceed. of the 26th Annual Meet. of the Assoc. for Computational Linguistics*; Buffalo, pp. 95-103.

Holmes, J. 1988: "Paying compliments: A sex-preferential politeness strategy," *Journal of Pragmatics*, 12, 4 (Aug. 1988), pp. 445-65.

Jackendoff, R.S., 1983: *Semantics and Cognition*; Cambridge, Mass.: MIT Press.

Johnson-Laird, P.N., 1983: *Mental Models: Toward a cognitive science of language, inference, and consciousness*; Cambridge: C. Univ. Press.

Johnson-Laird, P.N., 1986: "Reasoning with Logic," in Myers, Brown and McGonigle (1986), ch. 1.

Just, M.A., and P. A. Carpenter (eds.), 1977: *Cognitive Processes in Comprehension*; Hillsdale: Erlbaum.

Kaplan, R., 1972: "Augmented Transition Networks as Psychological Models of Sentence Comprehension," *Artificial Intelligence*, 3 (1972), pp. 77-100.

Kaplan, R., 1974: "On Process Models for Sentence Analysis," in D.A. Norman and D. J. Rumelhart (eds.), *Explorations in Cognition*; San Francisco: Freeman, 1975, ch. 5.

Kempson, R.M., 1986: "Definite NP's and Context-Dependency: A Unified Theory of Anaphora," in Myers, Brown and McGonigle (1986), ch. 9.

Kempson, R.M., 1988a: "On the grammar-cognition interface: the principle of full interpretation," in Kempson (1988b), ch.9.

Kempson, R.M. (ed.), 1988b: *Mental representations: The interface between language and reality*; Cambridge: C. Univ. Press.

Kimball, J., 1973: "Seven principles of surface structure parsing in natural language," *Cognition*, 2, 1 (1973), pp. 15-47.

Kimball, J., 1975: "Predictive analysis and Over-the-Top Parsing," in J. Kimball (ed.) *Syntax and Semantics*, vol. 4; New York: Academic Press, pp. 155-79.

Koike, D.A., 1988: "Requests and the role of deixis in politeness," *Journal of Pragmatics*, 13, 2 (Apr. 1989), pp. 187-202.

Kurcz, I., G.W. Shugar and J.H. Danks (eds.), 1985: *Knowledge and language*; Amsterdam: North-Holland, 1986.

Lakoff, G., 1969: "The Role of Deduction in Grammar," in C.J. Fillmore and D.T. Langendoen (eds.), *Studies in Linguistic Semantics*; New York: Holt, Rinehart & Winston, 1971.

Lakoff, G. and H. Thompson, 1975: "Introducing cognitive grammar," in C. Cogen *et al.*, (eds.), *Proceedings of the First Annual Meeting of the Berkeley Linguistics Society*; Berkeley: B. Ling. Soc., pp. 295-313.

Lapointe, S.G., 1983: "Some issues in the linguistic description of agrammatism," *Cognition*, 14, 1 (1983), pp. 1-39.

Leech, G., 1982: *Principles of Pragmatics*; London: Longman, 1983.

Levelt, W.J.M., 1983: "Monitoring and self-repair in speech," *Cognition*, 14, 3 (1983), pp. 185-209.

Lewis, D. 1974: "Radical Interpretation," *Synthese* 23 (1974), pp. 331-44.

Marcus, M.S., 1977: *A Theory of Syntactic Recognition for Natural Language*, Ph.D. dissertation, M.I.T.; published with slight revisions, 1979: *A Theory of Syntactic Recognition for Natural Language*; Cambridge, Mass.: MIT Press, 1980.

Marslen-Wilson, W., 1975: "The limited compatibility of linguistic and perceptual explanations," in R.E. Grossman *et al.* (eds.), *Papers from the Parasession of Functionalism, Chicago Linguistic Society*; Chicago: Ch. Ling. Soc., 1975, pp. 409-20.

Marslen-Wilson, W., 1976: "Linguistic Descriptions and Psychological Assumptions in the Study of Sentence Perception," in R.J. Wales and E. Walker (eds.), *Language Mechanisms: A Collection of Psycholinguistic Stydies*; Amsterdam: North-Holland, pp. 203-29.

Matsumoto, Y., 1988: "Reexamination of the universality of face: Politeness phenomena in Japanese," *Journal of Pragmatics*, 12, 4 (Aug. 1988), pp. 403-26.

McNeill, D., 1977: *The Conceptual Basis of Language*; Hillsdale: Erlbaum, 1979.

McNeill, D., 1987: *Psycholinguistics. A New Approach;* New York: Harper & Row.

Mehler, J., E.C.T. Walker and M.F. Garrett (eds.), 1981: *Perspectives on Mental Representation*; Cambridge, Mass.: MIT Press, 1982.

Minsky, M., 1974: "A Framework Representing Knowledge," in P. Winston (ed.), *The Psychology of Computer Vision*; New York: McGraw-Hill, 1975, pp. 211-77.

Motley, M.T., 1985: "Slips of the Tongue," *Scientific American*, 253, 3 (Sept. 1985), pp. 114-9.

Myers, T., K. Brown and B. McGonigle (eds.), 1986: *Reasoning and Discourse Processes*; Orlando: Academic Press.

Nadel, L., L.A. Cooper, P. Culicover, and R.M. Harnish (eds.), 1989: *Neural Connections, Mental Computations*; Cambridge, Mass.: MIT Press.

[Ochs-]Keenan, E., 1975: "On the Universality of Conversational Postulates"; Mimeographed by Indiana Univ. Linguistics Club.

Osgood, C.E., 1979: *Lectures on Language Performance*; New York: Springer, 1980.

Pinker, S., and J. Mehler (eds.), 1988: *Connectionism and symbol systems* [=*Cognition*, 28, 1-2 (March 1988)].

Sadock, J.M., 1972: *Toward a Linguistic Theory of Speech Acts*; New York: Academic Press, 1974.

Sadock, J.M., 1986: "Remarks on the paper by Deirdre Wilson and Dan Sperber," in A.M. Farley and K.E. McCullough (eds.), *Papers on the Parasession on Pragmatics and Grammatical Theory, CLS* 22, Part 2; Chicago: Ch. Linguistic Society, pp. 84-90.

Searle, J.R., 1974: "Indirect Speech Acts," in Cole y Morgan (1975), pp. 59-82.

Searle, J.R., 1979: *Expression and Meaning: Studies in the Theory of Speech Acts*; Cambridge: C. Univ. Press.

Searle, J.R., 1983: *Intentionality: An essay in the philosophy of mind*; Cambridge: C. Univ. Press.

Smith, N.V. (ed.), 1981: *Mutual Knowledge*; New York: Academic Press, 1982.

Sperber, D. and D. Wilson, 1981a: "On Defining 'Relevance,'" unpublished manuscript.

Sperber, D. and D. Wilson, 1981b: "Mutual Knowledge and Relevance in Theories of Comprehension," in Smith (1981), ch. 2, pp. 61-85.

Sperber, D. and D. Wilson, 1981c: "Reply to Gazdar and Good," in Smith (1981), ch. 2, "Comments and Replies," pp. 101-110.

Sperber, D. and D. Wilson, 1986: *Relevance: Communication and Cognition*; Oxford: Blackwell, 1986.

Swinney, D.A., 1981: "The Structure and Time-Course of Information Interaction During Speech Comprehension: Lexical Segmentation, Access, and Interpretation," in Mehler, Walker and Garrett (1981), ch. 8.

Thagard, P., 1984: "Frames, Knowledge, and Understanding," *Synthese*, 61, 2 (November 1984), pp. 233-59.

Vicher, A. and D. Sankoff, 1988: "The emergent syntax of pre-sentential turn openings," *Journal of Pragmatics*, 13, 1 (Febr. 1989), pp. 81-97.

Von Savigny, E., 1988: *The Social foundations of Meaning*; Berlin: Springer.

Wanner, E., 1980: "The ATN and the Sausage Machine: which is baloney?" *Cognition*, 8 (1980), pp. 209-25.

Wason, P.C., 1959: "The processing of positive and negative information," *Quarterly Journal of Experimental Psychology*, 11 (1959), pp. 92-107.

Wason, P.C., 1961: "REsponse to affirmative and negative binary statemente," *British Journal of Psychology*, 52 (1961), pp. 133-42.

Wilensky, R., 1986: "Some Problems and Proposals for Knowledge Representation"; Berkeley: Inst. of Cognitive Studies, Univ. of California at B.

Wilson, D. and D. Sperber, 1980a: "On Finding the Context for Comprehension," paper delivered to the European Psycholinguistics Ass. Meeting, Goteborg, Sweden.

Wilson, D. and D. Sperber, 1980b: "On Grice's Theory of Conversation," in P.N. Werth (ed.), *Conversation and Discourse*; London: Croom Helm, 1981, ch. 8.

Wilson, D. and D. Sperber, 1986: "Inference and implicature," in C. Travis (ed.), *Meaning and Interpretation*; Oxford: Blackwell, ch. 2.

Wilson, D. and D. Sperber, 1988: "Representtion and Relevance," in Kempson (1988b), ch. 6.

Winograd, T., 1972: *Understanding Natural Language* [=*Cognitive Psychology*, 3, 1 (1972)]; also as a separate publ., New York: Academic Press.

Winograd, T., 1987: "Thinking Machines: Can There Be? Are We?" [=Report No. CSLI-87-100]; Stanford: Center for the Study of Language and Information.

Woods, W., 1970: "Transition Network Grammar for Natural Langauge Analysis," *Communications of the Association for Computing Machinery*, 13, 10 (1970), pp. 591-602.

CHAPTER 7

PARTIALITY AND COHERENCE IN CONCEPT COMBINATION

Nick Braisby*
Bradley Franks
Terry Myers
Centre for Cognitive Science
University of Edinburgh
Scotland

I. INTRODUCTION

This paper addresses the problems of lexical representation and coherence. Our discussion of these problems leads to what we call a sense generation approach. We will present two accounts from within this approach and illustrate them with respect to concept combination. In section 2 of this paper we describe the tension that exists between the two phenomena of semantic flexibility and specificity. There are accounts of lexical representation in which this tension raises major problems and there are others in which it provides a clue to resolving these problems. The former are characterised by what Clark (1983) calls the selection of senses and the latter by what he calls sense creation. In section 3 we discuss the implications of sense selection accounts for coherence and concept combination, indicating problems that arise. We rely heavily on Murphy & Medin's (1985) arguments and suggest some extensions. In section 4 we present the sense generation view (which is related to Clark's notion of sense creation) which avoids such difficulties. Central to this account is a consideration of partiality, reduction of partiality being brought about by constraints provided by the discourse and situational factors. In section 5 we will sketch two different approaches to concept combination within this framework. In section 6 we draw conclusions and raise questions regarding the nature of coherence and lexical representation.

II. FLEXIBILITY AND SPECIFICITY

Consider the following example:

> Rudy is at the ballet watching a single ballerina dancing on stage. Excitedly, he whispers to his friends, "Isn't she delightful?" His friend Ron agrees, replying, "Yes, she's very beautiful." His other friend, Reg, agrees too, saying, "Yes, what an exquisite pirouette." But Rudy replies, "I simply meant she's a wonderful person."

Delightful in the above exchange seems to mean different things for the various individuals involved. From Ron's perspective, *delightful* means "physicallydelightful"; from Reg's, *delightful* means "dancing delightfully," while for Rudy *delightful* means "delightful as a person." From the perspective of an usher walking past who only hears Rudy's first comment, the meaning of *delightful* is indeterminate with respect to these specific meanings.

Suppose that we want to characterize the semantic component of the lexical entry for *delightful*. We might propose that the lexical entry is sufficiently general to encompass all the observed meanings. In which case, we require a lexical entry which is general enough to capture all the different meanings that can be conveyed by *delightful*, including those illustrated in the exchange. Even if we could specify such a general lexical entry for *delightful*, however, this would not then allow the possibility for *delightful* to convey, say, the meaning "delightful as a person." Yet, in use, *delightful* can convey just this. What is more, any of the meanings indicated above could be made even more specific if more contextually provided information were available. For example, Ron's meaning for *delightful*, "physicallydelightful" could have been more specific: "physically delightful with respect to coiffure" if we know that ron is, say, a hairdresser. In principle, there is no bound to this degree of specificity and so a general lexical entry seems even more implausible. Tied to this idea of unboundedness, we also want to allow that meanings can be indeterminate: for example, the meaning for the usher, is non-specific with respect to the meanings for Rudy and his friends. In conclusion then, our treatment of the lexical entry as general, being motivated by the observation of flexibility, renders the lexical entry unable to provide for the specificity of meaning that we observe. Seemingly, we can't treat the phenomena of flexibility and specificity independently of one another: to do so produces the tension we have illustrated.[1]

III. SENSE SELECTION

3.1 Weak and Strong Sense Selection

For the purposes of a distinction to be made later, we will refer to the semantic contribution of the lexical entry as the "lexical concept," and the intended meaning as the "generated sense." One approach to characterising lexical entries is to treat the different meanings of a word as different lexical concepts. For example, under this view we might suppose that the lexical entry for *mother* includes several lexical concepts for *mother*, which we might label "foster mother," "step mother," and "biological mother"—in the same way as, say, *bank* is assumed to have a number of independent lexical concepts underlying its use. A strong version of this view would hold that for each and every different sense a different lexical concept is accessed from the lexical entry. A weaker version maintains that there may be independent, unrelated lexical concepts from which, on different occasions of use, different senses may be elaborated. For example, under the weak view, we could, in the context of a fairy tale, arrive at a sense for *mother* corresponding to "step-mother who is wicked" by elaborating the lexical concept for *mother* labelled "step-mother." Presumably, on the strong sense selection view we would select the appropriate sense, "step-mother who is wicked," relevant to the context and that could not be further elaborated.

Both versions of sense selection miss the crucial intuition that these senses are closely related; they are not at all like the different senses of *bank*. That is, *river bank* and *money bank* have clearly distinct meanings, and thus *bank* should have (at least) two different lexical concepts. In the case of *mother* the different lexical concepts we could hypothesise are clearly related. Accordingly, in the strong version, we have a granular picture of flexibility and specificity. Flexibility is granular in that the possible senses a word may have cannot alter and we have an impoverished view of flexibility. Specificity is also fixed by the set of senses in that senses can be no less or no more specific than those represented. The weak version allows senses to vary in terms of their specificity but we only have apparent flexibility: the range of possible senses is restricted by the initial choice of lexical concept.

3.2 Coherence in Sense Selection

In exploring the implications for coherence of the sense selection view we will treat Lakoff's (1987) analysis of *mother* within this account, although we accept that he may not actually be committed to such a view. That is, we might see each of his sub-models as a different lexical concept within the lexical entry for *mother*. A number of consequences then follow. In particular, we will argue that we cannot maintain both weak sense selection and a theory-based view of coherence. To see this we must turn to Murphy & Medin.

In Murphy & Medin's (1985) account we may distinguish between internal and external components of conceptual coherence. Both aspects derive from theories of the world. Internal coherence results from the relations between features for which the theory provides an explanation. For example, the various properties of chairs represented in the concept (its size, having a seat, its rigidity and so on) cohere in virtue of theories concerning the function of a chair. External relations between different concepts are also motivated by theories that support causal and explanatory connections between those concepts. Consider the concepts "cat" and "dog." According to Murphy & Medin these concepts are coherent to the extent that they are embedded in our theories of the world: the greater the number of connections between such concepts and the rest of our knowledge, the more coherent and stable they are. The connection of "cat" to many different aspects of our knowledge (such as our theories of domestic life, ownership, pleasure and so on) provides it with coherence over and above that which is provided by our theories of "cat" alone, which is the source of the concept's internal coherence. The same thing is true of complex concepts, for example "cat and dog." The concept is coherent as a result of there being many common theories of cats and dogs. Seemingly, on Murphy & Medin's view, we could not allow that complex concepts such as "elephants and lemons" or "stone lions and trout" have the same level of coherence as "cat and dog." This stems from the fact that such complex concepts has fewer common theories to relate the component concepts.

This view of the theories underlying coherence seems to be intimately connected to metaphysical issues. Accordingly, for Murphy and Medin, concepts result from the way that theories cut up the world. Consequently, categories that cut across ontological boundaries would not be very coherent. For example, "elephants and hopes" would not be considered coherent, unless this category were motivated by a theory. If this theory were consistent with one's other theories, the coherence of the concept would be enhanced. The claim of Murphy & Medin is that the flexibility of conceptual coherence derives from the flexibility of particular theories. On our interpretation of their claim, this means that there are two mechanisms for allowing coherence to be flexible. Either, in certain contexts, coherence is enhanced by constructing an explanation or theory or, alternatively, the theories that underpin coherence are inherently flexible. It is not clear how these mechanisms for flexibility in coherence would work for examples like the following:

> Imagine you are sitting on a park bench in London. In front of you is a stone statue of a lion. Consider how you would reply, if someone were to ask you, "Is that a lion?" Seemingly, there is one sense in which you might say "yes," and a related one in which you might say "no": and, given these senses, either reply could be appropriate, according to circumstances.

It seems, then, that the word *lion* can be used quite appropriately to talk of a statue of a lion. However, there is a question as to just how many of our theories of lions involve statues. Seemingly, our theories of lions are capable of cutting the world up so that we can categorically assert that a statue of a lion is not a lion. One approach to this phenomenon is to view it as a peculiarity of work use. However, we wish to preserve the intuition that there is an essential perspectival aspect to this case. That is, we can adopt a different perspective on a statue of a lion: we may simply view it as a lion. Additionally, in this case each of these perspectives seems equally coherent. As we see it, there are two possibilities for Murphy & Medin to account for this apparent flexibility in coherence. One is to invoke their suggestion regarding the flexible nature of theories. Another is to adopt the standpoint of weak sense selection, and thus hold that there are two lexical concepts for *lion*. In what follows we will explore the former possibility and conclude that the latter is their only plausible option. This in turn leads to an inconsistency.

Under the first option, that of invoking flexibility in theories, we envisage three possible mechanisms open to Murphy & Medin for achieving the flexibility of coherence. One mechanism that Murphy & Medin offer is that of the construction of an explanation in cases such as "elephants and hopes" where, seemingly, ontological boundaries are crossed. In the above case, apparently there is a similar crossing of ontological boundaries. Our theories of lions tell us that stone lions are not really lions, though presumably there are some theories of lions that are also theories of stone lions. Perhaps then it is that the coherence of the concept of lion in this case relies on a constructed explanation. However, since our stable, underlying theories concerning lions cannot all apply in the case of the stone lion, it must be that the use of lion in talking of a stone lion is less coherent than the use of lion in talking of a real lion. This follows because coherence in Murphy & Medin's view falls out of the number of explanatory links which structure the concept. Hence, on this view such constructed explanations necessarily lead to less coherent concepts than those stable explanations provided by the knowledge base. This is at odds with our intuitions. Seemingly both uses are, in this case, equally coherent. What we want is to allow that both *ad hoc* and stable explanations can provide for equal coherence. This is what we will argue for in section 4.2.

Murphy & Medin provide another mechanism to account for the flexibility in coherence that we observe. This is the possibility that we may have flexible theories. In the case of the stone lion, we might suppose that our theories concerning lions are flexible, accounting for the observed flexibility of coherence in use. The use of *lion* to talk of a stone lion can be coherent as we would want in virtue of the fact that the same theories concerning lions can also be theories concerning stone lions. However, this does not accord with the intimate connection that theories have to metaphysical concerns. If our theories concerning lions are also theories concerning stone lions then we have no principled manner of cutting the world such as to differentiate between non-lions (like our stone lion) and real lions. We take it that this undermines the metaphysical position that Murphy & Medin adopt.

A third possibility that we believe Murphy & Medin may allow for is the option of theory change. This option entails that in order to preserve the coherence of the lion concept when employed to talk of a stone lion, the theories underlying the concept would have to change. This, however, seems problematic. Since they adopt the Quinian position of viewing theories as components of a large and intricate web, a change in one will have ramifications for the others. In this case we would have to change our theories concerning lions so as to allow the possibility of inanimate lions. Changes as radical as this, as they acknowledge, would require a global reorganization of one's knowledge base. It would also require, under most interpretations, a divorcing of theories from metaphysics. This is precisely what we argue for. In Section 5 we will present two views that are not tied to the assumptions that lead to these counter-intuitive consequences. As we have seen, there are good reasons to suppose the kind of flexibility observed in coherence is not to be captured by the notion of flexibility in theories. We now turn to the second possibility we pointed to earlier.

The alternative position that we believe Murphy & Medin might adopt is the one that we call weak sense selection. That is, the above arguments will not go through if we assume that there are two lexical entries and hence two lexical concepts for *lion*. Aside from the arguments already presented against this option, it seems there are further considerations that would prohibit Murphy & Medin from taking this view. Following Quine (1960), we might like to distinguish between two lexical entries for *light*. This move seems legitimate. After all, it seems that the theories we have for light-weight objects do not have to be the same as those for light-coloured objects. For the case of *lion*, although lions and stone lions are differentiated by the theories provided by the knowledge base, lions and stone lions are nonetheless very closely related.

It is presumably the case then that since we can relate lions and stone lions we have theories to do just this.

Returning to our interpretation of Lakoff, we can see that similar arguments apply. That is, if we tie Murphy & Medin's arguments to this assumed position then we may allow that the different sub-models for *mother* are related by what they call theories. Consider the following example:

> Kim lives with her biological mother, Mary. Mary adopts a child, Jane, to whom Mary behaves in all respects as she does to Kim. Jane calls Mary "mother" and behaves towards her in the same way Kim does. Initially, Kim cannot accept that Mary is Jane's mother but after some time she does.

How is it that Kim's notion of mother appears to change? One possibility is that initially she only has the "biological mother" concept of mother and that she somehow acquires another concept of mother, the "adoptive mother" concept. Even though Kim starts off with the notion of mothers as biological mothers, one can envisage that Kim's theories of the world allow her to construct an explanation of how Jane can call Mary "mother." Indeed, she may come up with a new notion of mother, a mother who is like a biological mother in all respects save the biological relations. That is, Kim, with her notion of a biological mother, equipped only with her theories of the world, can construct the notion of an adoptive mother. The fact that this seems so clearly to be the case undermines the starting point of weak sense selection. Namely, that there may be independent lexical concepts for *mother*. The fact that they may be related by our theories of the world indicates that they are not independent.

We have seen that to posit independent lexical entries is to undermine those very arguments that Murphy & Medin advance in favour of theories. Yet, we take those arguments as very good grounds to accept the theory-based account of concepts. Consequently, we also take it that what is needed to avoid the pitfalls of the sense selection accounts is an extension of this theory-based view. So, although we crucially need a way of describing the difference between lions and stone lions and mothers and adoptive mothers, we need a way of avoiding making an unprincipled distinction at the level of lexical entries.

There are several possible responses to this difficulty. One is to place constraints on the way theories may relate these independent lexical concepts. That is, we may want to place constraints on the possible extensions of any two of the lexical concepts for *mother*, such that one cannot be extended to form the other. However, this flies in the face of the observation that we can do just this. Kim *can* construct a notion of adoptive mother from her notion of biological mothers. So for this to be a viable option, more justification is needed. Another option is simply to reject the arguments of Murphy & Medin that the coherence of senses is determined by theories. The fact that biological and adoptive mothers may form a coherent complex concept has nothing to do with the theories that relate them. As stated earlier, we agree with Murphy & Medin's general position. We believe that the only other alternative is to reject the assumptions on which weak sense selection is based. As a consequence, we can see that the interpretation we placed on Lakoff in order to discuss coherence must be rejected. Accordingly, we can only view Lakoff's sub-models as describing something other than lexical concepts.

IV. SENSE GENERATION

As we have shown, we can characterise lexical entries according to the assumptions of sense selection. We believe that strong sense selection does not provide a viable characterisation. We have illustrated weak sense selection by choosing to interpret Lakoff in this manner. This position is challenged by the arguments in favour of theories as determinants of coherence. In this section we will present an alternative account of lexical representation that is consistent with such arguments.

4.1 Partiality

From our consideration of the arguments of Murphy & Medin concerning coherence, we require independent lexical concepts only in those cases where it is clear that they are unrelated by the kind of theories they discuss. For example, in the case of *bank*, we would require two lexical concepts, one for *river bank* and one for *money bank*. These, after all, seem unrelated unless we construct some *ad hoc* explanation. In the case of *mother*, though, we would not want to hypothesise independent lexical concepts for "foster mother," "step-mother," etc. Instead we would want to represent *mother* with a single lexical concept. In order for this concept to apply in the cases of all types of mother, that is to capture the phenomenon of semantic flexibility, we would seemingly require it to be general with respect to these cases. As we indicated before, this appears to disallow the possibility of capturing the phenomenon of specificity. However, this conclusion is tied to the assumption that lexical concepts and senses are total objects. Under this assumption, the sense for *mother* cannot be extended or elaborated. That is, we can only have one sense for *mother*. However, that there *are* many different senses is undeniable.

An alternative is to view the various senses as extensions of a single lexical concept, a view that is related to that of Dunbar (1988). That is, senses are related to lexical concepts via a relation of concept extension.[2] Returning to our example of Rudy at the ballet, the different meanings we may associate with *delightful* can be ordered in the following way. The usher's meaning is only as specific as the lexical concept, and this meaning is compatible with any of the possible extensions of the lexical concept. Ron's meaning, however, is more specific. It doesn't just indicate that there is something that is delightful, it also indicates what it is that is delightful, namely the dancer. Thus, for Ron, the information conveyed by *delightful* would appear to contain the information gleaned by the usher. Similarly, for Reg and Rudy with respect to the usher. Thus, the ordering is one of informational containment. However, if we compare the information conveyed by *delightful* from Ron's perspective with that conveyed from Reg's perspective, we cannot say that either is more informative. Thus the relation of informational containment is undefined for these two senses. It follows that the relation of informational containment is a partial one.[3]

Our position then is that semantic flexibility can only be properly captured by considering the lexical concept as a partial object. This can be differently extended according to the interaction of various constraints on each occasion of use. We see such constraints as being provided by the discourse, the situation in which it resides and cognitive models which pertain to both the discourse and the situation. It will emerge that these are the very constraints that underpin the generation of a coherent sense. That being the case, we will see that coherence and flexibility are but two sides of the same coin.

4.2 Coherence in Sense Generation

We have so far considered the coherence of concepts from the point of view of Murphy & Medin's arguments. On their view, coherence is intimately bound to metaphysical issues. As we have seen there are a number of problems associated with this approach. In what follows we will be concerned with the coherence not of concepts but of senses. In this way we avoid being committed to a necessary connection between coherence and metaphysics.

Taking lexical concepts as partial objects seems to provide us with too many degrees of freedom. So, for instance, we seem to require a constraint to prevent certain co-occurrences of features. Consider our lexical concept for *balloon*. We might want to ascribe to it the property of not being able to float in the air. Under the unconstrained account of the extension of partial objects, this is perfectly permissible. However, following on from the arguments of Murphy & Medin, we would want to claim that such an ascription could only be valid if licensed by a theory. In this case, the balloon could be filled with a heavy gas. Hence, in order to constrain what extensions we allow, the process of extending a lexical concept to form a sense must be the very process by which that sense becomes coherent. That is, the process of extension must realise the kinds of theory envisaged by Murphy & Medin.

In order to illustrate the kinds of issues which emerge with respect to coherence in sense generation, we will consider the example of the stone lion, of section 3.2. According to our view, a given sense of a word derives from the corresponding lexical concept and may or may not be an extension of it. Consider describing a stone lion as a lion. We can account for this sense of lion in either of two ways. One possibility is that the lexical concept is general and no extension is required. From our preceding arguments this option would disallow the possibility of simultaneously satisfying the conditions of semantic flexibility and specificity. The only option then is to consider that the lexical concept is specific in the following sense: it contains, or may inherit, among others, feature descriptions such as "animate," "four-legged" and "golden-colour." This will capture the sort of specificity we require for some occasions for use, for example, where we are describing a lion at the zoo. There may, however, be circumstances in which a lexical concept would be extended further, perhaps in order to distinguish between lions and lionesses. More importantly though, as it stands, the lexical concept cannot meet the flexibility requirement. In order to resolve these problems, we will invoke a more general mechanism for concept extension. So, in the case of the stone lion being described as a lion, seemingly the feature "animate" present in the lexical concept is denied in the generated sense. Accordingly, what we require of the constraints which underpin concept extension is that they allow this sort of defeasibility. In section 5 the two views we present will adopt different methods of achieving this.

In principle, we allow that any property ascription provided by a lexical concept can be defeated. As a consequence, we cannot regard the various entities described by a given word as forming a category which has some metaphysical significance. That is, such categories, in contrast to those envisaged by Murphy and Medin, do not "carve the world at its joints." Returning to the example, since a stone lion is inanimate, we require that the extension of "lion" involves at least the defeat of the property ascription concerning animacy. In this way we can meet the requirement of semantic flexibility. Given the lexical concept we have for *lion*, there are circumstances in which it seemingly can't be used. However, via our defeasible mechanism for concept extension, we can override any feature which would prohibit its use. The important point here is that this overriding can only be licensed by theories, *ad hoc* or otherwise.

The mechanism as we have described it so far is a little simplistic. Consider again the case of the stone lion being talked of as a lion. In one sense, we may agree that the statue is a lion but in another we may protest that it is not. That is, we require some choice in the mechanism as spelled out. If we want to affirm that the statue is a lion, then we will adopt the defeasible method of extension already outlined.

If, however, we want to deny that the statue is a lion, we will not extend the lexical concept in this defeasible way. Of course, we may allow the statue to be described as just that, a statue. Again, in such a case we wouldn't require any defeasible extension of the lexical concept for statue.

These different options for concept extension lead to different property ascriptions, that is, different descriptions concerning the same entity. We take these descriptions of an entity to define different perspectives. For a particular cognitive agent, a perspective associates descriptions with given entities. So, if Fred approaches our stone lion adopting the "real lion" perspective, one that is defined by the description given by the lexical concept for lion, he won't categorise it as a lion. However, were he to adopt the "statue of a lion" perspective, one given by a description resulting from defeasible extension of the lexical concept for lion, he would categorise it as a lion. He could of course adopt a third perspective, the "stone statue" perspective, resulting from a non-defeasible extension of the lexical concept for statue. From this perspective, he would similarly categorise the stone lion as a statue.

We can now see that an adequate characterisation of coherence is not going to be provided solely by the theories which Murphy & Medin discuss. Instead, we will pursue the idea that theories may be local: that is, provided by *ad hoc* explanation, situational factors and informational requirements. If Fred were to exclaim, "This is not a lion," then we can only ascribe to him the perspective we labelled the "real lion" perspective. It is only this perspective that makes this use of lion coherent. If we attribute the "statue of a lion" perspective to Fred, then this particular use of lion would be rendered incoherent. It follows then that the coherence of a sense can only be defined relative to some perspective. So, clearly, perspectives may involve local or *ad hoc* theories: we don't carry around our theoretical luggage everywhere we go.

There are many constraints on concept extension that give rise to coherence. Some are to do with the extra-linguistic aspects of word use, such as the agents involved, the situations in which they find themselves and the particular informational requirements they have. Others are to do with the linguistic context, concept combination, for example. In what follows we will focus on the notion of informational requirement and examples of concept combination.

Consider the following puzzle taken from Braisby (1989):

> Fred is sitting on a park bench in London. He knows that at the other end of the park there is a statue of a lion. A schoolgirl approaches him and, explaining that she has been given an assignment to sketch a lion, she asks Fred if he has seen one. Fred replies that he has and points her towards the statue. A little later an exhausted zoo-keeper appears and, explaining to Fred that a lion has escaped from the zoo, he also asks Fred if he has seen one. Fred replies that he hasn't.

Related to our discussion of perspectives, we would like to say that Fred is adopting different perspectives for different occasions. The reasons for choosing the particular perspective he does are to do with the nature of the information required of him. Fred can infer that the schoolgirl requires information germane to lions that one can draw and so adopts the relevant perspective in order to reply in the way that he does. The zoo-keeper, however, places a different informational requirement on Fred. It is this that similarly helps determine the perspective Fred adopts and so allows him to reply as he does to the zoo-keeper.

We would like to distinguish two aspects to informational requirement. One is appropriate to the degree of specificity a given sense may have. So, in the above example, in replying to the zoo-keeper, Fred means "real lion" by *lion*, a meaning given by the lexical concept for lion. In replying to the schoolgirl, however, Fred has a more specific meaning in mind, namely *lion* as meaning "statue of a lion," a meaning that can be derived via defeasible extension from the lexical concept for lion. That is, the informational requirements of the agents involved help determine the degree of specificity of the sense of Fred's use of *lion*. In addition, these requirements lead to a difference in the flexibility of these meanings. That is, the meaning of Fred's use of *lion* in replying to the schoolgirl, is only flexible enough to describe statues of lions. The meaning of his use of "lion" in replying to the zoo-keeper, however, is different in terms of flexibility: it is only flexible to the degree that it can describe real lions. So the informational requirements of the agents involved can also lead to differences in the flexibility of meaning.

The other constraint on concept extension that we will consider is that provided by concept combination. In the next section we will expound on this within the general framework developed so far. We will do so via the application of two alternative views.

V. PARTIALITY, COHERENCE AND CONCEPT COMBINATION

Our preceding arguments lead to a set of conclusions that we take as requirements on any theory of lexical representation. First and foremost is the commitment to sense generation, in which non-ambiguous words are represented by a single lexical entry, which contributes one lexical concept, and particular senses are generated from this by concept extension. Further, we take the lexical concept to contain a description of the central properties of entities which can be described by the corresponding word. Accordingly, the lexical concept is not general in the Quinian sense; that is, in itself it can only describe a subset of the range of entities which the word itself can describe. Quine's notion of generality would maintain that the lexical concept can describe all individuals the word is used to describe. Entities outside of this subset, but which nonetheless can be described by the word, require a different mechanism. The mechanism we propose is one of concept extension. We see this as being determined by theories, though not ones that carry metaphysical assumptions, *a la* Murphy & Medin. As a consequence, the extensions of concepts are senses, whose coherence is provided by the nature of the extension. The kinds of theories which underpin the generation of senses are those that allow for defeasibility. Similar to the "lion" case we discussed earlier, we will treat such extensions as defining perspectives, and accordingly coherence obtains relative to a perspective. Perspectives themselves are constrained by factors of the situation, local context and the informational requirements of the agents involved. The following two views are based upon these conclusions.

5.1 View 1

This section outlines a possibility for lexical representation developed by Braisby (1989). Braisby refers both to an implementation in C-prolog and also a situation-theoretic interpretation of this implementation. For purposes of brevity, we will not go into these matters here. What we will present is necessarily a simplification. However, the main points will still emerge. In that account a word meaning (or WORM) can be described in a similar way to that offered by the standard classical view. For example, the WORM for *lion* could be described as follows:

$$\begin{bmatrix} \text{animacy:animate} \\ \text{colour:golden} \\ \text{legs:4} \end{bmatrix} \qquad (1)$$

When Fred responds to the zoo-keeper, what he is claiming is that he hasn't seen an individual which this description satisfies. But this WORM does not help us in the case of the schoolgirl. For Fred has not seen *any* individual which can be described by this meaning. How then can he reply to the schoolgirl that he has seen a lion? In Braisby's (1989) account, this case implicates what are called Combinations of sc worms (or COWORMs). COWORMs can also be involved in word meaning. Indeed, a word's meaning may be given by the WORM which corresponds to it or by any one of a number of COWORMs. A COWORM is arrived at by combining a number of WORMs. Taking the "lion" case as an example, there is a combination of the WORMs for *lion and* statue as follows.

WORM(LION, P_1) (2)
WORM(STATUE, P_1 , P_2)

What the COWORM does is to relate various data structures. It does this by virtue of the fact that "statue" relates two types of data structures. That is, we may view it as an operator. Given a data structure of one particular type, we obtain a data structure of a different type. For example, this particular COWORM, relates the data structure above to one like the following:

$$\begin{bmatrix} \text{animacy:inanimate} \\ \text{colour:\_} \\ \text{legs:4} \end{bmatrix} \qquad (3)$$

Here the "statue" WORM is being used to form a different kind of data structure. In particular, it tells us that a statue of a lion is inanimate and does not inherit the colour it has from the thing it represents. The data structure given by this COWORM describes perfectly what Fred *has* seen. Thus, Fred's reply to the schoolgirl is based on this (or a related) COWORM. The crucial aspect of this account is that it is COWORMs which involve a given WORM that may underly the meaning of that WORM's corresponding word. It is given that this is a condition on word use. The particular COWORM that is adopted by a speaker or hearer depends on a large number of factors, ones that have been discussed in the earlier parts of this paper. It is clear in this case though that it is the nature of the information sought from Fred that determines the particular WORMs and COWORMs he employs. This is not unlike the case of Rudy.

Let us suppose that each of our excited ballet lovers has the same WORM for the ballerina as follows:

$$\begin{bmatrix} \text{body:} \begin{bmatrix} \text{shape:fine} \end{bmatrix} \\ \text{hair:} \begin{bmatrix} \text{colour:brown} \end{bmatrix} \\ \text{dance:} \begin{bmatrix} \text{move:pirouette} \end{bmatrix} \\ \text{personality:} \begin{bmatrix} \text{temperament:quiet} \end{bmatrix} \\ \vdots \end{bmatrix} \tag{4}$$

Here a subtlety in the process of combining WORMs is required. When "is delightful" is predicated of the ballerina by Rudy, his friends may do a number of things with the WORM that describes her. One option, that chosen by Reg, is to combine his WORM for the ballerina with that for *delightful* so the feature of delightfulness is added to that aspect of the WORM for ballerina concerning her dance as follows:

$$\begin{bmatrix} \text{body:} \begin{bmatrix} \text{shape:fine} \end{bmatrix} \\ \text{hair:} \begin{bmatrix} \text{colour:brown} \end{bmatrix} \\ \text{dance:} \begin{bmatrix} \text{move:pirouette} \\ \text{is:delightful} \end{bmatrix} \\ \text{personality:} \begin{bmatrix} \text{temperament:quiet} \end{bmatrix} \\ \cdot \\ \cdot \\ \cdot \end{bmatrix} \tag{5}$$

In a similar fashion, Ron combines his WORMs for the ballerina and *delightful*, resulting in a data structure in which "delightful" is a feature of the body attribute of the WORM for the ballerina. Similarly for Rudy. How, though, are we to capture the usher's interpretation?

The usher has a meaning for the ballerina that is the same as our ballet lovers, let us say. However, she has many possibilities open to her. She may combine her WORM of *delightful* with that for the ballerina in any number of ways. The number of options open to her are determined in two ways. Firstly, by the number of complex attributes of the WORM for the ballerina. Secondly, by the number of additional WORMs that she could employ. Similar to the "lion" case where Fred used the additional WORM of "statue" according to the appropriateness of the circumstances, so the usher has similar options too. If, for example, the usher is aware that certain members of the audience play all manner of language-games, then the meaning of *delightful* is simply indeterminate. That is, there are many possible COWORMs that we may take to define these games and the usher has no way of deciding which COWORM to choose. It is this indeterminacy that renders her unable to say what Rudy means.

Our other example involves Kim, Mary and Jane and the problems Kim has when Jane comes to stay. In particular, we want to say that Kim has a notion of what it is to be a biological mother but not an adoptive mother. However, she must come to acquire such a notion. The crucial aspect of the general discussion has been that we do not want to conclude that Kim ends up with more than one lexical entry for *mother*. Again, we will use the mechanisms of COWORMs.

We can describe Kim's WORM for *mother* as follows.

$$\begin{bmatrix} \text{gender:female} \\ \text{genetics:gen} \\ \text{child: [genetics:gen]} \end{bmatrix} \tag{6}$$

How is it that given this WORM for *mother* Kim can later accept Jane's utterances concerning Mary? Again, we will claim that Kim learns of "adopting," which expresses a relation that Kim can utilise in forming the following COWORM.

$$\begin{array}{l} \text{WORM(MOTHER, } P_1\text{)} \\ \text{WORM(ADOPTIVE, } P_1, P_2\text{)} \end{array} \tag{7}$$

This COWORM will give a new kind of data structure, of a type related by the WORM for *adoptive*. This particular WORM acts to deny the genetic connection between parents and children. Accordingly, we get the following data structure which allows Kim to make sense of Jane's use of *mother*.

$$\begin{bmatrix} \text{gender:female} \\ \text{genetics:gen1} \\ \text{child: [genetics:gen2]} \end{bmatrix} \tag{8}$$

Within Braisby's framework WORMs and COWORMs can be taken as defining different perspectives. Thus, we might say that Fred has two perspectives: one where he uses the WORM "lion" in isolation, one where he uses it in combination with the WORM for *statue*. Similarly, what Kim has learned is not a new lexical concept or WORM for "mother" but a new way of Combining WORMs (or a new perspective) and similarly with our long forgotten ballet lovers. In a similar vein, we may envisage "concept combinations" being accommodated in this fashion. Let us take as an example, "stone lion." It seems that we would like to say that, at the very least, "stone" implies inanimate while "lion" implies animate. Such an intuitively appealing view precludes the standard set intersective account. Alternatively, the feature structures which describe the component WORMs will fail to unify. Here, we would expect one or the other of the component WORMs to be extended so that unification will succeed between the extended senses. Just as before, a Combination is required. Combining the WORMs for "lion"and "statue" renders a sense for *lion* that implies inanimacy. The feature structure described by this COWORM can then unify with that for "stone." However, just as we may view this as taking a particular perspective on what "stone lion" means, we can imagine that this combination allows for a number of different perspectives to be taken. So, although we may take the "statue of a lion" perspective defined by the COWORM of "lion"and "statue," we could also take the "petrified lion" perspective on "stone lion" which is defined by the combined WORMs of "lion"and "petrified."Given suitable intonation and stress, we might also adopt the perspective which gives the meaning "lion that sells stones" which is what we would want to say in the case of "coal merchant" which means "one who sells coal." And so on.

We have seen that by considering the meaning of a word as being described by different data structures on different occasions of use, we open up the possibility of capturing flexibility and specificity. In addition, coherence is captured by the use of COWORMs which we may think of as embodying atomic, local theories. This is consistent with the objectives of the sense generation position.

5.2 View 2

This section presents some of the central aspects of the view of lexical representation and combination sketched in Franks et. al., (1988), and developed in Franks (1989). The outline given here is a simplification of that work in that it will focus primarily upon concept combination. We will consider the way in which one concept's extension constrains that of the other in a head-modifier relationship. In this way, the combined extensions result in a single sense for the noun phrase. The properties descriptions of the lexical concepts correspond to "Central" properties, derived from conventionalised lay theories. These include Central-essence properties, which relate to the ontological or functional essence of an entity.

Let us again consider our ballet buffs. Suppose that the ballerina is known to them as "Netty."Suppose further that Rudy's first comment had been:

(1) Netty is a delightful ballet-dancer.

Each of the meanings of Rudy's comment, for the ballet-buffs, can be seen to interpret (1) in different ways. On a simple account, we might say that this combination involves the concept for *ballet-dancer* being extended to incorporate the property of delightful. Hence, we might then say that the interpretation of (1) is (2).

(2) Netty dances ballet delightfully.

This is Reg's interpretation. It is based upon an assumption of Explicit Semantic Attachment, in which the major source of the properties of the sense for a NP is the explicitly given head noun, and the properties contributed by the modifier will be evaluated with respect to this noun (Platts, 1979). In (1), the explicitly attached noun is "ballet-dancer." Franks argues that, in order to capture semantic flexibilitywe must allow for Implicit Semantic Attachment: that is, agents may evaluate the property of being delightful with respect to a different noun concept. How is this effected? The initial combination of the modifier with the head noun unifies their properties and results in a sense which carries the meaning that Netty, who is a ballet-dancer, is delightful in some unspecified way. This is the sense which is generated by the usher. Her sense is compatible with many different extensions, in which the general property of being delightful is determined as a more specific property by evaluating it with respect to an implicitly attached noun concept. For example, Ron's meaning involves a choice of a noun concept such as "people seen in terms of their bodies," which results in an appropriate specification of being delightful. We then have this sense for (1):

(3) Netty is a ballet-dancer who is delightful to look at.

And Rudy's sense for (1) will similarly involve the selection of another implicitly attached N concept. In these cases the different agents are adopting different perspectives on the ballerina. We noted in 4.2 that a perspective defines a description of an entity. Franks argues that perspectives have two components: the categorising sortal noun or nominal phrase (either implicitly or explicitly attached to the construction which describes the entity), and the range of properties which are used to make the categorisation. In this case, we have perspectives whose major contrast is given by their different covering sortals. In contrast, the perspective of the usher has a less specific range of properties, and no particular sortal noun. Consequently, for the usher, the meaning of Rudy's phrase is indeterminate. The flexibility in perspectives will be further evident in the consideration of *stone lion* and similar combinations.

If combining concepts constrain the extension of the head noun, then they may act as defeaters for some of the properties in the noun's lexical concept. This is exactly what happens in what Franks labels Negating Functional Privative combinations, such as *stone lion* and *wooden banana*. These behave in a similar way to Negating Privatives Proper, which involve adjectives such as *fake* and *false*. The latter have usually been defined as denying such inferences as that from

(4) This is a fake gun.

to

(5) This is a gun.

Franks argues that this conclusion omits the crucially perspectival nature of such categorisations, which we will now examine.

Consider the combination "fake gun." The lexical concept for *gun* might include such Central-essence properties as "internal mechanism for propelling bullets" and "barrel for directing bullets," and other properties such as "trigger mechanism," "barrel mounted over trigger," "colour and weight of metal," and so forth. The lexical concept for *fake* specifies that it will defeat the Central-essence properties of "gun," and allow the ascription of some of the "appearance" properties of "gun"to the sense for *fake gun*. The modifier, then, acts as a Rebutting defeater for the noun's Central-essence properties, that is we can ascribe the negation of the properties to the sense. After this combination, where the modifier has priority, we find that the sense for *fake gun* parallels the usher's for *delightful ballet-dancer*. That is, it includes only the general property that the entity has some of the appearance properties of the head noun. But we cannot say in precisely which way it looks like a real gun. It is consistent with many different extensions of this "appearance" property, through the selection of a particular implicitly attached noun. This is consistent with our intuition that there are many possible ways for a gun to be fake. An entity of the type described by this sense would be indiscernible from a range of entities of the type of possible implicitly attached noun concepts, which might be labelled "toy gun," "replica gun" or "model gun." This means that we might adopt a categorising perspective on such an entity based on the narrow range of "appearance" properties, with respect to any of these nouns, as well as the head noun. That is, we might categorise it thus:

(6) The fake gun is a gun (or toy/replica/model gun) with respect to appearance.

Since this perspective does not incorporate any Central-essence properties, it could not be generalized to a broader range of properties, and hence the categorization is restricted in its applicability. We will refer to such restricted perspectives as Type II Perspectives.

If there is an informational requirement to make the sense more specific, then our agent will access an implicitly attached noun in order to extend the sense. This choice is crucially related to situational factors. If we know that the entity described by *fake gun* has been used to overcome some intruder in a house, we might then select "replica" as the appropriate noun type. This will be used to specify the detail of the "appearance" property, and the sense will also inherit the Central-essence properties of this noun. So the sense for *fake gun* includes properties to the effect that such an entity is a replica of a gun, which "looks like" a real gun in the way that a replica does. Now, we might adopt two perspectives on the entity described by the noun phrase. The first is the Type II Perspective:

(7) The fake gun is a gun with respect to appearance (as a replica).

The second is a Type I Perspective:

(8) The fake gun is a replica with respect to Central-essence properties.

A Type I Perspective is one in which the categorized entity has the Central-essence properties of the categorising sortal noun.

It seems that exactly the same kind of process occurs for a combination like *stone lion*. Here, the sense can define two different perspectives. Fred's situational context would facilitate a choice of "statue" rather than "rock-formation" or "ornament" as the implicitly attached noun. The Type II Perspective is:

(9) The stone lion is a lion with respect to shape (as a statue).

This is the perspective of the schoolgirl, which Fred adopts in order to respond as he does. The Type I Perspective is:

(10) The stone lion is a statue with respect to Central-essence properties.

The perspective of the zoo-keeper would be a Type I: to satisfy his informational requirements, the entity described by *lion* would need to be a lion with respect to Central-essence properties. This clearly cannot be supported by the sense generated for *stone lion*. In order for Fred to reply "yes"to any agent asking about the presence of lions, the perspective of the questioner must incorporate a range of categorising properties which subsumes the range of properties in the sense for *stone lion*. The schoolgirl's perspective satisfies this, but the zoo-keeper's does not.

Franks argues that this is a requirement on the coherent use of a categorisation statement.

Our third example concerns Kim's quandary. This might be approached through a consideration of another type of privative combination, the Equivocating Privatives. These include combinations such as *blue orange* and *straight banana*, as well as ones involving adjectives like *apparent* and *alleged*. Equivocators and combinations in which we are unable to state whether the entity described by the noun phrase is in fact a member of the head noun or not. We simply have insufficient information. The modifier Undercuts the head N's Central-essence properties. That is, on current information, we are unable to ascribe these properties to the sense: it does not mean, unlike in the case of Negators, that we can ascribe the negation of these properties to the sense. If our information changed, we might reverse the Undercutting.

Consider the combination "apparent friend." The Central-essence properties of the concept for *friend* might include "loyalty,""affection," "absence of self-seeking," whilst the other properties include "solicitous behaviour," "smiles,""various vague kindnesses." And the Modifier's lexical concept specifies the Undercutting of the head noun's Central-essence properties, and an ascription to the noun phrase sense of some of the non-central-essence properties. Having Undercut the Central-essence properties of the head noun, we are again left with a sense which is indeterminate as regards exactly how the "appearance" properties of the head noun are manifest. After accessing an appropriate implicitly attached noun, such as *deceiver*, the general property is made more specific. This sense then defines a Type II Perspective on an entity described by the noun phrase:

(11) This apparent friend is a friend (or deceiver) with respect to appearance.

In order to decide whether this apparent friend is a real friend, we need to extend our information through some process of inquiry. The outcome of this will determine the appropriate combinations of perspectives. In an Affirmative outcome, the apparent friend is a real friend, thus reversing the Undercutting of the Central-essence properties of the head noun, and allowing for a Type I Perspective:

(12) This apparent friend is a friend with respect to Central-essence properties.

In a Privative outcome, the sense defines the same combination of perspectives as in the Negating Privatives case.

Let us return to Kim's quandary. This follows a pattern similar to Equivocators. In trying to understand Jane's use of the word *mother*, through trying to adopt Jane's perspective, Kim will Undercut the Central-essence properties of her concept of *mother*. Hence she could use this sense to define Jane's perspective, which is Type II:

(13) Mary is Jane's mother with respect to behaviour.

This allows her to have some understanding of how Jane can call Mary "mother." However, Kim may learn of "adopting," which enables her to determine that, from Jane's perspective, the sense for *mother* involves a Rebuttal of the Central-essence properties of the lexical concept for *mother*. But as in the Negators case, this does not prevent Jane from calling Mary "mother": it simply means that she must adopt a particular type of perspective to do so. As in View 1, this implies that Kim is not learning a new lexical concept for *mother*, as in our interpretation of Lakoff, but that she is able to adopt a new perspective—a Type II Perspective—on her mother as Jane's mother.

On this view, then, the flexibility and specificity of a wide range of combinations are addressed by the twin mechanisms of defeasible extension and implicit semantic attachment. Both are viewed within the context of the perspective-relative nature of categorisation. Accordingly, coherence is relative to the two aspects of a perspective, and both must be appropriate for a coherent categorisation.

VI. CONCLUSIONS

In this paper, we have outlined a general account of lexical representation and presented two approaches within this account and the way they treat the problems of coherence and concept combination. Moreover, we have demonstrated that both approaches promise an adequate treatment of the crucial phenomena of semantic flexibility and specificity. These approaches have been motivated by arguments concerning coherence in sense selection.

Strong sense selection has unfortunate consequences especially regarding the combinatorial explosion of interpretation associated with complex expressions. We take it that this position is not only unfortunate but untenable. Weak sense selection, which is assumed in a number of linguistic approaches, avoids an explosion of this magnitude. However, we have argued that the position is undermined by a consideration of coherence. We concur with the general position of Murphy & Medin that coherence is determined by theories. As a consequence, we expect the relations between senses to be expressed in any theory of lexical representation. This conclusion is in direct opposition to the starting position of weak sense selection. Our conclusion is that such a starting position is unprincipled. In order to have a principled account of the phenomena we have discussed, it seems a sense generation account is required.

An important implication of the sense generation approach is that lexical concepts should be regarded as partial. The specificity of meaning we need is brought about by the possibilities of concept extension. The mechanism of concept extension, in allowing for the defeat of properties, also provides for the desired flexibility of meaning. Importantly, the constraints on extension are provided by theories which, as we have seen, underpin the coherence of the generated sense. We have strongly argued that theories are local: they may be provided by the immediate situation. In addition, agents can choose to utilise certain theories in preference to others on the basis of their informational requirements. This at once extricates us from Murphy & Medin's connection between metaphysics and theories. It also, crucially, leads us to the conclusion that an agent adopting different theories in his cognitive activities is entertaining different perspectives. Consequently, the examples which we have employed throughout this paper are essentially perspectival in nature as, indeed, is word meaning.

NOTES

*The order of authorship is completely arbitrary. The last two authors were supported by European Basic Research Action, ESPRIT project 3175. The first two authors were supported by ESRC studentships Nos. C00428722001and C00428622024respectively. We would like to express our gratitude to Richard Cooper, Robin Cooper, George Dunbar and other members of the Mental Lexicon Workshop for many useful discussions and insights.

1. A similar point has been made by, among others, George Dunbar. In Dunbar (1988) he suggests that, "while lexical meaning is flexible, lexical use can be very precise...this is paradoxical" (p. 69).

2. From here on we use *concept extension* to mean that relation between a lexical concept and the same lexical concept some of whose features have been modified. Modification here includes the cases of simple adding of features as well as the denial of features. In no sense do we mean *extension* to refer to the set of objects described by the lexical concept.

3. Since the relation is partial, it follows that concept extension need not result in a sense that is more specific than the lexical concept.

REFERENCES

Braisby, Nick, 1989. Situating Word Meaning. Paper presented at the Conference on Situation Theory and its Applications, Asilomar, CA.

Clark, Herbert H., 1983. Making Sense of Nonce Sense. In d'Arcais, G.B. Flores and Ray J. Jarvella (eds.). *The Process of Language Understanding*. Chichester: John Wiley and Sons, 297-331.

Dunbar, George L., 1988. The Cognitive lexicon. Ph.D. Thesis. Centre for Cognitive Science, University of Edinburgh.

Franks, Bradley, 1989. Concept Combination: Towards an Account of Privatives. Edinburgh Research Paper in Cognitive Science.

Franks, Bradley, Terry Myers and Scott McGlashan, 1988. Defeasibility in Concept Combination. In *Proceedings of the Tenth Annual Conference of the American Cognitive Science Society*, Montreal.

Lakoff, George, 1987. Cognitive Models and prototype theory. In Neisser, U. (ed.). *Concepts and Conceptual Development: Ecological and Intellectual Factors in Categorization*, 63-100. Cambridge: Cambridge University Press.

Murphy, Gregory L. and Douglas L. Medin, 1985. The role of theories in conceptual coherence. *Psychological Review*, 92, 289-316.

Platts, Michael, 1979. *Ways of Meaning*. London: Routledge and Kegan Paul.

Quine, Willard V.O., 1960. *Word and Object*. Cambridge, Mass.: MIT.

CHAPTER 8

THE LABYRINTH OF ATTITUDE REPORTS

Daniel Quesada
Universidad Autónoma de Barcelona
Spain

In this paper I attempt once more to solve the thorny issue of the interaction of semantic and pragmatic aspects in the interpretation of attitude reports.[1] I will here approach the subject in a sort of (relatively) "naive" way. In particular, I will not go at all into the technical aspects of the proposals that will be put forward. I will be concerned mainly with reports that use sentences having as grammatical subjects of the subordinates either definite descriptions or other kinds of noun phrases, such as noun phrases with possessive pronouns, personal pronouns and demonstratives.

Before I begin the discussion of the topic, I have to outline, even if in a very dogmatic and crude form, a perspective on which mental states are, which will constitute the background of the views on attitude reports to be explained below. Furthermore, I need to at least barely indicate the framework of semantic ideas in which the present paper is placed, that is, I need to say something about semantic contents. Both needs are satisfied in the first two section of the paper.

As will be seen, I adopt the view that it is essential for mental states to have semantic content. In the main bulk of the paper, I turn to the relation between this content and the content of the embedded sentence in a report about such mental states. In the third section I criticize a first proposal that could be made on that relation. Yet another proposal, the one being advanced currently by Crimmins and Perry, is explained in the following section, presenting at the end of the section the sort of case with which I anticipate trouble for the account.

What I hesitatingly take to be the weakness of this very interesting proposal is pointed out in section six, after presenting a distinction of my own between two kinds of attitude reports, a distinction which I definitely think must be taken into account in thinking about such reports. The distinction is further related to the inclusion of attitude reports among folk psychology explanations of behavior, which is made in section five. This identification seems to square well with one kind of report, but also seems to present some difficulties with the other kind. I try to make some steps toward resolving these difficulties by giving some further details of my proposal in the final section.

I. MENTAL STATES[2]

Very roughly put, mental states are physical "internal" states (brain states)—and, as such, causally linked with other states, external input and behavior—, which are semantically described.

Mental states are relevant where we are interested in explaining behavior of entities that cannot be satisfactorily explained as a more or less direct reaction to the environment. Thus, *pace* purely introspective theories of mental states, all their essential properties must be derived from an analysis of their connections with such behaviors. And we can get at the properties that are of interest to us here in (very roughly) the following way.

The complexity of the movements of certain beings (among which, of course, we are included) makes us think of such movements as internally caused. Thus, we postulate *internal states* as causes of the movements. A second central ingredient of such movements is that they are "successful" or not, that they lead to situations of the beings in their environments which are *adequate* or *inadequate*.

Moreover, when considering a certain subclass of such beings, we are forced to attribute to them some kind of "registering" the adequacy or inadequacy of their own behaviors. Since what they do next depends largely on whether what they did was successful or not, this plasticity is found only where the behavior is *learned* rather than being instinctive or previously designed. And it seems that we are ready to apply the label "mental states" for the internal states which are causes of movements only when these further conditions ("registering" of adequacy/inadequacy, and learning) are met, and thus perhaps only properly to human beings and other higher animals—at least for the time being.

What we see then in connection to the organisms to which we attribute mental states is that they *strive to* do something adequately. The states causing their behaviors "propel" them to modify their relation vis-à-vis their immediate surroundings in such a way that they reach an outcome which is adequate *because* it is adequate, even if they sometimes, or even often, fail. We thus sort out those states in, at least, two kinds: the states that are correlated with the adequate situation to be reached, and the states correlated with the situations in which the being actually exists.

The former states have directly to do with the adequacy of the situation to be reached after the behavior (or *action*, as we more specifically are by now ready to call the relevant kind of behavior): they determine whether this situation is adequate, in the sense that for the situation to be adequate it has to somehow "adjust" to the state. We call these mental states *conative states* (wishing or desiring is an stereotypical conative state).

But, in their own way, the second kind of state has also very much to do with the adequacy or inadequacy of the states: normally, the behavior or action has a chance to be successful only if such states "adjust" to the situations the animal is in, and so, by the process of learning, the organism "enlists" as causes of movements only those among them that produce behaviors leading to situations that are registered as adequate. These states we call *cognitive states* (or also *doxastic states*, taking belief as the stereotype).

In this way, the notion of adequacy of an action leads to attribute to mental states *semantic properties* as something that is in the nature of such states. Conative states are satisfied or not by a situation, the situation reached by the organism in question after the action "prompted" by those states. Cognitive or doxastic states are made correct or incorrect, true or false by the situations the being is in. And, if they have such semantic properties, they must have *semantic content*, that is, associated propositions, satisfaction conditions, truth conditions, or whatever we decide is the best way to "materialize" semantic contents.

Now, the naturalist view of the world inclines us to attribute the property of being a cause of something only to physical states, and thus, those mental states must be also physical states, but physical states semantically described as they must be if we are to understand how they cause the actions they do cause.

It would be in principle possible to try for a position according to which all the above line of thought is based on the possibilities of understanding action given our present state of ignorance, and that we will in due time, with the advancement of science—be it the computational or the neurophysiological studies of processes, or a combination of the two—forget completely about mental states, or semantic descriptions of physical states causally responsible for behavior. I think this position is wrong, and that the "mental state" view—the semantic descriptions of cognitive states and processes—is indispensable, the basic reason for this being perhaps that one and the same state or process semantically described can be instantiated by several different cognitive states or processes functionally or neurophysiologically described that satisfy equally well the semantic description.[3]

Certainly, in a paper dedicated to the interpretation of the ordinary language sentences or utterances by which we attribute mental states to agents, it is hardly to be expected that I go into this question. On the other hand, it cannot be denied that the overall relevance of the issues raised by the attempts at such a semantics is seen in a different light when one thinks that such attempts concern uniquely common idioms to be completely put aside by scientific talk as compared with thinking that this is to basically misjudge the situation.

Be this as it may, what is most important for present concerns is to stick to our conclusion that to have a semantic content is in the nature of a mental state, conative, doxastic, or other, if there it may be.

II. SEMANTIC CONTENTS

To barely indicate the framework of semantic ideas in which the present paper is placed, the following points about the semantics of natural language expressions have to be recorded. Of course, they are about expressions as they are used in declarative sentences not occurring in embedded sentences in an attitude report.

1) The main component of the semantic content of a *name* is the referent. In particular, names do not have a content that could be given with a definite description that either would be synonymous with the name or would fix its reference. The reference of a name is determined contextually in the speaker's subscribing a denominative practice.[4]

2) For *definite descriptions*, the main thing to bear in mind is the difference between their referential use and their attributive use. This distinction is explained differently by different authors, but for present purposes one can just formulate it by saying that the content of an utterance of a declarative sentence, with a definite description as its subject being used referentially, is a singular proposition, whereas the content of the same sentence, when the description is used attributively, is a general proposition.[5]

3) For *personal pronouns* and *demonstratives* what we need is the difference between their meaning when we make abstraction of context of use (e.g. "I" means: the speaker of the utterance) and the object they contributed to the proposition expressed in each context of use.[6] This is what I call here the content of those expressions.

When we move from semantics to psychosemantics, that is, when we move from considering contents of natural language expressions to contents of mental states or thoughts, we are stepping on even more difficult terrain. The main relevant conclusion here is that psychosemantics must be much more "fine-grained" than semantics. That is, the objects considered to be contents of thoughts or mental states must be less extended in space or time than the objects regularly picked up as referents of natural language expressions.[7]

III. ATTITUDE REPORTS AS EXPLANATIONS

An attitude is just a mental state, and as such individualized by its kind and its content. It just happens that it is so called—"attitude"—because it is named after its kind.[8] An attitude report is then just a report about a mental state.

In giving a theory of attitude reports, the main difficulty lies in making a systematic proposal on the content of the embedded sentence used in the report that is adequately related to the content of the mental state such a report is about.[9] A first proposal that could be made is based in the following principle:

(P1) The semantic content of an embedded sentence as used in a true attitude report must be identical to the semantic content of the attitude the report is about.

There are different ways of interpreting and implementing this schematic principle. (P1) could be seen, in principle, either as constraining the things that can be contents of embedded sentences, or the other way around, as constraining the things that can be contents of mental states. The perspective on mental states endorsed above leads one most naturally to the first possibility. One could then, for example, say that, besides their conventional contribution to the proposition expressed by the sentence —in a certain context of use, semantically relevant expressions have a special, usually different content when used to describe mental states, and then, that *this* is precisely the contribution they make to the propositional content of the sentence when it appears embedded as part of a larger sentence which is used for making an attitude report. The "purpose" of such a shifting in semantic content of expressions in attitude reports would be, of course, to keep track of the semantic contents of the states attributed.

Postulating semantic shifts has been the mark of a Fregean position on the semantics of attitude reports. Suppose now that we round off the just described (possible) semantic position by saying that it accepts such conclusions about the conventional semantic contributions of names, descriptions and indexicals in "non embedded" uses as are found in the work of Kripke, Donnellan, Kaplan, and Perry. To the extent in which there is still a shift in the postulated semantic content of expressions that depends on expressions appearing in certain linguistic contexts, we should perhaps still classify this semantic position as "semantically neoFregean."

Since I do not want at this point to go into exegetic details, I will not attribute this position to anyone. Still, the "semantically neoFregean" I have in mind, actual or merely possible, is also "cognitively neoRussellian" in that she thinks that mental states are characterized by semantic content, and this in turn by objects and properties in the world.

The immediate difficulty that such a sort of "synthesis position" seems to face is that it must maintain that expressions are semantically unsuspectedly "fine grained" in attitude report contexts, if they are to match the "fine-grainness" of the semantic contents of the attitudes themselves, while, on the other hand, the content a sentence is assigned in a context of use by the semantic conventions of ordinary language will hardly be prone to such a degree of "fine-grainness." Ordinary language is an all-purpose public means of communication, and, as such, it would hardly be suited to carry very detailed information about concrete, particular states of individuals.

I think that it is this mismatch which contributes most to the "labyrinthical" character of the subject of attitude reports. It, for example, poses an immediate difficulty for the sort of "synthesis position" I have briefly outlined above, since such a view is bound to maintain that expressions in attitude report contexts can be unsuspectedly "fine-grained" semantically, if they are to match the "fine-grainness" of the semantic contents of the attitudes themselves.

Thus, we can see that there are serious difficulties with a position that directly embodies a version of principle (PI), endorsing—in the described manner—the view that an attitude report is about the semantic content of the attitude.

IV. THE CRIMMINS-PERRY THEORY

A theory that tries to travel far without a resort to "semantic shifts" is the one advanced recently by Mark Crimmins and John Perry.[10] It also accepts (PI), at least for a bunch of central cases. But it avoids the shifts for those cases by seeing the principle as constraining the semantic contents of the *states*, not of the embedded sentences in context. That is, the states the report is about, have, in the relevant cases, the very same content as the conventional usual semantic content of the embedded sentence as used in the report.

Now, it would seem that no-one could do so without giving a pragmatic account of well known substitutivity problems. That is, it would seem that nobody could maintain that

(1) Miles Hendon did not believe that he was of royal blood,

and

(2) Miles Hendon believed that Edward Tudor was of royal blood.

are both *true* (assuming now, for the sake of the example, that it does not concern any fictional character), and assert at the same time that "he" in report (1) refers to the same object that "Edward Tudor" refers in (2). But Crimmins and Perry succeed in doing precisely that. Let us then pause to lay out the basics of their proposal, and show how the mentioned consequences follow from it.

According to the Crimmins-Perry theory, a belief report is an existential claim to the effect that there is a belief meeting a certain condition and with an specified semantic content. The key novel element in the proposal is that it is part of the semantic content of the report as a whole that the belief in question meets the condition.

The belief itself is a particular belonging to an agent at a certain time, and the condition a property of such particulars. One could then imagine the belief as a structured particular, having constituents that could be called *ideas* (ideas of properties or *property-ideas* and ideas of individuals or *notions*).

Taking this picture into the theory of the reports, the proposal is then (for the simplest sort of cases) that a belief report is an existential claim to the effect that there is such a belief of an agent at a certain time, having a certain semantic content and meeting certain a condition that concerns the belief and its constituent notions. In this way, the report claims something *about the notions* of the agent, and thus, the mere existence of a belief with certain content does not suffice for the truth of the belief report.

Let's consider an example. Suppose I talk about John Perry to my students, saying he is a Stanford Philosophy professor, notorious for his work on the interaction of language and mind, and the co-author of a well known book, *Situations and Attitudes*. Later on, instructing someone interested in what my students know about Perry, I report:

They believe that Perry teaches Philosophy at Stanford University.

The students had—from my explanation—picked up a *notion of* John Perry, and my present report is (partially) *about that notion*. It claims that there is a belief my students share, whose content is the proposition that John Perry teaches Philosophy at Stanford University, *and* that the notion is responsible (in the belief in question) for who fills the argument role of the property of teaching philosophy at Stanford. This last condition is full of technical or quasi-technical terms, but it amounts to, roughly, a condition about how, according to the report, my students "conceive" of John Perry. Thus the report *claims* something about the way in which the agent conceives objects. That is, then, a necessary part of the total claim, and the report is not true if that part does not hold.

But where does that part of the claim come from? To start with, where do the notion or notions the report is about come from? They come from the context. It is the context of the report which makes clear which is the notion or "way of conceiving" the object by the agent (or which, at least, constrains the notion in some way). Such notions and conditions are "*unarticulated constituents*" of the report content as a whole: contexts bring them in, not anything in the words used.[11] But such "ways of conceiving" do not infest the content of the belief itself. According to the report—linguistic utterance *plus* context—the agent's belief gets that content *through* such and such notion or notions, so to speak, but the content itself remains uncontaminated by any notions, ideas or "ways of conceiving."

In this way, we reach a reassuring state of affairs. The semantic contribution of the embedded sentence in the belief report to the semantic content of the report itself is just the semantic content of its utterance in the context. There is no "shift of interpretation" for embedding. Some version of "semantic innocence" is still preserved. To that important extent the semantics is not neoFregean.

On the other hand, we still get something that could be called "ways of conceiving" in the picture. However, they—we are assured—do not come this time from the embedded sentence, but, somewhat unexpectedly, from the context. And nevertheless, they have a role to play in determining the semantic content (the proposition or truth-conditions) of the whole report.

It seems we have here a nice "division of labor": the embedded sentence takes care, in the context, of the semantic content of the belief; the context, all by itself, of the cognitive perspective of the agent.

With such an account one sees easily how one can go then to resolving the doxastic puzzles. The key is, of course, that an agent can have some "combination of ideas" with some specified semantic content without having some *other* "combination of ideas" that would have the same content. A report which would state the first would then be consistent with a report that denied the agent has the second belief or "combination of ideas"—as it is exactly the case with (2) and (1)—and a report with the same wording that some other, but issued in a different context could make a different claim, since it would probably be about different notions (this takes care of the famous Kripke's Pierre puzzle, as we will see in a moment).

However, in spite of all its virtues, I see trouble for the account. It makes the attitude reports completely centered in the perspective of the agent, and thus we should indeed expect trouble coming from reports that do not bring in that *specific* perspective. And I maintain there are and always have been such reports. Here is an example.

A and B see Ralph Running in pursuit of Bernard J. Ortcutt. Let us suppose, for making the issue clearer, that somehow both A and B know that Ralph is not in a position to know that the man he runs after is called "Ortcutt," and that each one knows that the other knows just *that*. Now, in that situation, B asks to A: "Why is Ralph running after Ortcutt?". To which A answers back issuing the following report: "He believes Ortcutt to be a spy."

Examples do not bring by themselves much conviction. But I will below put this kind of example in a more comprehensive context in order to advance my own view of attitude reports.

V. REPORTS AND REPORTING

One important consequence of the view of Crimmins and Perry—one not at all affected by possible shortcomings it might have—is that focusing on mental state *reports* rather than in the *sentences* used, allows one to dismantle the *assumptions* in which some of the most known (so-called) doxastic puzzles are based. Take on most privileged example: the famous Pierre puzzle, due to Kripke. This puzzle is based on two apparently reasonable assumptions:

(A1) If a competent speaker of a language assents, reflectively and sincerely to "*p*," then he believes that *p*.

(A2) If a sentence of one language expresses a truth in that language, then any (correct) translation of it into another language also expresses a truth.[12]

As Crimmins and Perry explain,[13] these assumptions do not any longer carry weight when considering *reports*, because the same sentence can be used to make true and false reports.

We conclude, then, that focusing in the reports removes some important difficulties. But we cannot at all leave the issue there, because it is clear that the removing of the difficulties hardly explains why we (must) focus on the reports in the first place. I think that we must take next *reporting* itself into account, recognizing thus the simple fact that a report is the result of an activity done by somebody, with some purpose. and, stepping on this path, we come most naturally to the question: which is the primary purpose of such a mental state reporting?

I want to subscribe here to the most straightforward answer, and this is, it seems to me, that the reporting is done primarily with the aim of explaining some piece of behavior. That is, in reporting about the attitudes, the (primary) purpose of the reporter, the speaker, is to *explain* the behavior of the agent to some other person(s), the "audience." In saying this is most straightforward, I rely, of course, on the fact that, as it was briefly explained above, the nature of mental states themselves makes them causes of actions. And thus, a report about such causes may be fit for an explanatory function.

Now, in which way exactly is the speaker trying to explain an action when issuing an attitude report, since she in informing about the mental states of the agent—to be sure the ones that (partially, at least) caused the action—is describing those internal states semantically. Mental states, we agreed, are internal states semantically described or classified, that is, internal states described by their semantic content. The speaker is then, according to the present view, causally explaining an action, but describing the relevant cause only semantically.

A sole report is only about one state of the agent, while for the explanation of action we will require in general more than just one state. Usually then, the explanation connected with the report is highly elliptical, relying heavily on context and background factors. When, for example, the report is a belief report, it is explanatory only against the background of some assumed goals or desires of the agent. Thus the report is claimed to be really only part of an explanation. Once, however, this has been acknowledged, explicit mention of it will be largely avoided in what follows, just for simplicity.

When we see reports as fulfilling an explanatory function, we are making of them part and parcel of a common sense or "folk"psychology, a psychology that, among other things supplies people with action explanation of a certain kind due only to the fact that they are persons, regardless of any special scientific competence they may have. And, if I am right, knowledge that mental states have "constitutively"a semantic content is part of such a "folk"psychology.[14]

When explaining an action by invoking an internal cause of it, it would seem that the cause so invoked must be internal to the agent. On the other hand, if our causal explanation is accurate, it seems it must pick out the right cause. In particular, since the cause—the internal state—is "semantically described," the description of the semantic content of it should be right. Thus, at first sight, it would seem as if making attitude reports explanatory has the immediate effect of forcing one to accept principle (P2):

(P2) The internal state which is semantically described in a true attitude report is the specific state of the agent whose attitude the report is about.

Moreover, since according to the present view, the speaker uses in a certain context a certain sentence for the description of the cause (the embedded sentence of the report), it would seem that this sentence should then, as used in the context, provide the content of the mental state. Thus, it would seem that such a view forces us to accept the version of principle (P1) I first described right after stating the principle.

This would not be a happy situation, since we have already found some difficulties in maintaining (P1), and, as we will also see, there are difficulties with (P2). Fortunately, the bearing of our present view about the primary function of attitude reports on principles (P1) and (P2) is significantly different that the one just described.

VI. TWO KINDS OF ATTITUDE REPORTS

I think that the clarification of the issue just raised at the end of the preceding section is closely related with the distinction between two kinds of attitude reports. This distinction is related to the old *de dicto/de re* distinction, but at the same time it is sufficiently different from it as to require some elaborate explanation. Let me first motivate the distinction by bringing in *three* little alternative stories. Common to the three is that two persons, say A and B, are chatting about some things happening around. Furthermore, in the first two, A utters at some time the sentence:

(3) Mary believes that John's lover is her best friend.

(*Story 1*) A and B have been talking gossip about Mary's husband, John, having a lover, without Mary having the faintest idea of the fact; now a points to a woman telling B: "she is John's lover." After a while, both A and B see Mary herself effusively greeting the woman. B asks A for an explanation of Mary's behavior ("Why is Mary behaving so?"), becoming finally (3) as an answer.

(*Story 2*) Similar gossip going on between A and B, and same finger pointing by A to John's lover. However, A says now that Mary is aware of John's infidelity, and that she also knows who the lover is, being in fact well acquainted with her. The same irruption of Mary in the scene and a similar effusive greeting take place. B, again, surprised by Mary's behavior, questions A, getting (3) as an answer.

(*Story 3*) A mix of stories 1 and 2. A says to B that Mary suspects John has a lover. Same woman on the stage, same coming into of Mary, same warm greeting, same question. This time, however, A says:

(4) Mary does not believe she is John's lover.

I think that the right grouping of these stories pulls the *first* apart from the remaining two. Here is, in broad outline, how I propose to analyze the examples.

In the third one, the contribution of "John's lover" to the content of the report is not just an individual. The context makes it quite clear that, from the perspective of Mary, there is no specific individual involved. Thus, "John's lover" is used attributively, as intimated roughly by the handy paraphrase "John's lover, whoever she may be," and therefore the content of the embedded sentence as used in the report is a general proposition.

In contrast to this, in the second story the contribution of "John's lover" to the content of the report is just an individual, that person on the stage, the description being used referentially. Thus, the content of the embedded sentence of the report is a singular proposition.

Thus, semantically considered, the two reports are clearly different in just this relevant respect: in the second report a singular proposition is involved, while in the third it is a general proposition that is at stake. What unites them however, is that in both reports we have some speaker trying to explain Mary's action from the perspective of Mary, so to speak. The background of the third report makes it clear that from her perspective there is no specific individual involved. But just so, the background of the second report makes it clear that there is for Mary a very definite individual involved. It is this similarity that we can capture saying that in both cases the speaker is (*relatively*) *specific* or that he is *attributing a (relatively) specific content* to the belief held by the agent. That is to say, the speaker succeeds in informing not only about the object concerned, but about more specific features of the belief, about how this content is entertained by the agent, hinting—if one wants to use that terminology—about the notions or ideas involved. This success is made possible largely by the context or background: the speaker relies upon it to get the additional information through.

Contrast all this with the first report. I think it will be readily admitted that the contribution of the description "John's lover" to the content of the report is just an individual. Thus, the semantic content of the embedded sentence as used in the report is again a singular proposition, just as in the second report. But in this case, I think that the background makes it clear that the speaker does not adopt the perspective of the agent. She knows full well that the agent has no idea of who John's lover is, and she further knows, by the time she issues the report, that B, the audience, knows that as well. Therefore, we should say that the speaker is *(relatively) unspecific about the belief*, classifying it only with the individual and property involved, and not giving additional information about its "composition."

I am thus proposing the following view. One can classify attitude reports in more than one way. One possibility is grouping beliefs by the kind of content attributed: singular or general. That would put together the reports of stories 1 and 2, separating them from the report of the third story. But I think that, at least as revealing as such classification, would be to group reports by "relative specificity" about the attitude attributed, in the manner I have just suggested. Thus, according to this classification, in a report of the first kind—illustrated by stories 2 and 3 above—the speaker is explaining the action "from the perspective" of the agent, and so, the words she uses constituting the embedded sentence have the purpose of describing the specific state of the agent that are causally responsible for the action. To this end, the speaker is using the context to suggest further details about the state, in addition to its classification by content. And so, we can say that the speaker attributes to the agent a belief with a content which is the content—singular or general—of the embedded sentence used in the report.

In the second kind of report, the perspective adopted is not the perspective of the agent. The speaker is not trying to capture with any specificity the state the agent may have. Unconcerned by this aim, the speaker is classifying the state in a too rough a manner for us to say that he is attributing to the agent a state with a content that corresponds to the content of the embedded sentence. Below I will try to turn this purely negative characterization into something a little more satisfactory.

Finally, it is part of the same view that common to both kinds of reports is that in the situations involved the speaker is explaining the action of the agent, and thus that they should be considered as parts of such explanations. That is why the speaker classifies—in all cases—the states *semantically*, by their content.

The drawing of the distinction is relevant as well for the appraisal of the Crimmins-Perry theory, because I think it does not accommodate it. In particular, I do not think that can be plausibly maintained that reports of the second kind—the ones illustrated by the first story or by the new Ortcutt example—impose any further condition on the belief reported or are *about* any notion of the agent, although, of course, if there are such things as notions, some notion or other will be involved in the state reported.

Let me illustrate the distinction again, at the same time showing further aspects of it. If you and I know who is John's lover, and we are discussing the natural color of her hair, I may say: "Eleanor is blond." But if you ask me why has Mary just said that she hates blond women, and I know that you think that she—although already suspecting or actually knowing that John has a lover—does not know who the lover is, I will say: "Mary thinks that John's lover is blond." Why I am likely to use the words "John's lover"? Because I, as a competent speaker of English,[15] know the expression to be a description, and I further know that descriptions can be used attributively, and that, when so used, the proposition expressed is a general proposition, and, finally, I know as well that you, as a competent speaker of English, know all this. This is the semantic part of it. But I had also a pragmatic reason for using those words. I thought that, given your knowledge of the situation, you would take the description as attributively used, and interpret the embedded sentence as expressing a general proposition, and, recognizing my intention, as finally attributing to Mary a belief state with a general content. If this is how you interpret it, everything went right, since it is really so how I wanted to be interpreted.

But suppose now that you pursue the conversation saying: "Ah! Has then she finally found out who John's lover is?" What would have gone wrong here? Well, you would then think that I was attributing to Mary a belief state with a singular content. In part, this means that you did not interpret me as saying: "Mary thinks that John's lover (whoever she may be) is blond" but as saying something of the following sort: "Mary thinks that she (Eleanor, that woman) is blond." You interpreted the description as referentially used. But what is important her to notice is that you would still think that I was describing things from Mary's perspective, and so being specific about her state.

Observe now that, in fact, I just could have been issuing such a report with the intention to attribute to Mary a belief state with a singular content. A is doing just that in Mary's story number 2 when issuing (3) to B. Indeed, he issues (3) knowing that the chances are that "John's lover" is taken to contribute just the referent, the other woman at the stage, to the content of the report, and thus to the description of the content of the state so reported. In fact, the success of what he intends to communicate depends on this to be the case. And he knows the chances are so, partially because he knows that "John's lover" can be used referentially, and knows that B knows that. And partially also, because the situation makes the relevant woman salient enough to provide a referent. It is then clear that, if in the case alluded to, you would interpret me as attributing to Mary a belief whose content is a singular proposition, the blunder would not be due to lack of semantic or pragmatic competence on your part.

But there are situations that make for the speaker very unreasonable to intend that she should be taken to be describing the agent "state of mind" with even that level of specificity. Such situations overlap greatly with contexts in which would be unreasonable for the hearer to take the speaker to intend that. The whole situation of story number 1 provides, I think, an example. And the new Ralph-Ortcutt story I made up at the beginning gives us still another. In these cases, to repeat myself, the speaker is really not intending to attribute to the agent a belief state which has as its content the content of the embedded sentence, even if, as in the first example, a singular proposition may be involved in the report. Indeed, for that matter, I do not think that the agent can be taken to attribute to the agent any specific state. He is just not trying to describe the agent's mind, not making any effort to "see" things from her perspective.

It would make some sense to call the first kind of report *de dicto*, and the second kind *de re*. But it must be clear just which distinction these words are made to label here.

With many others, I think that, somewhere around the subject "propositional attitudes," there is a significant difference between two kinds of things. I also share the opinion that the traditional labels should not be applied to the attitudes themselves (in a certain sense, all of them are "de re"). But what I am pressing especially here is the view that the difference is also not semantic. To begin with, it is clearly not just a difference in the meaning of the (embedded) sentences involved. The difference is in the reports. But, furthermore, it is not a difference in the semantic content of the reports which constitutes the relevant distinction; it is not that with one kind of report we attribute mental states that have a general content, while with the other kind we attribute states having a singular content. The same kind of content can be involved in reports of both sorts. The decisive point is whether the report purports to specifically describe "the agent's mind" or not. And it is an speaker which "does purport something" or not, that is, that intends the report to be taken in a way or the other. Thus, it is a *pragmatic* distinction that is here at stake.

A pretty good test to find out which kind of report is involved in a particular occasion is the "assent-test." When the report is *de dicto* (in the above sense), the agent knows the language and is ready to collaborate, there is every reason to expect—if the report get things right—that she would assent to the embedded sentence, if no personal pronoun occurs in it, or to a related sentence with the pronoun(s) appropriately changed, if such a pronoun does occur. And on the contrary, since *de re* reports are unconcerned about the perspective of the agent, there is no reason at all to expect the agent to assent to such a sentence when the report is of that kind.[16]

VII. REPORTING AND EXPLAINING

Until now, I have been describing the distinction in very broad terms as reports in which "things are seen from the perspective of the agent" or reports in which "things are not so seen." I think that some progress can be made in explaining these coined phrases by thinking of the activity of uttering sentences as a classificatory activity.[17] That is, uttering (declarative) sentences in certain contexts or situations amounts to classifying the world (or a portion thereof) in one way or the other.

Applying this point of view to attitude reports, we see that in uttering a certain sentence for reporting about an attitude of some agent, a sincere speaker is classifying "the world" as containing an agent having a mental state of a certain kind and with a certain content. This "high level" classification is entirely the responsibility of the speaker. But, when issuing a *de dicto* report, there is a component in the classification on which she, so to speak, "delegates responsibility" to the agent. And this is the content of the state reported. Mental states themselves classify (parts of) the world, and so the speaker tries to reflect that classification, the classification made by the state, the best she can. The best she can, that is, given common sense or "folk" psychology, the explanatory aim of the report, her linguistic competence, the context in which the report is produced, and assorted limitations she may have (time, memory, and so on).

That is why, if the speaker is correct about the state, she has got its content right; right, that is, relative to the coarseness of the classificatory schema involved. But it is coarse classificatory schemes that are involved in folk psychology explanations. They rely apparently on exploiting gross regularities holding among coarse types of situations. And this is how we can reconcile the thesis that an attitude report reports about the semantic content of an internal state with the alleged fine-grainness of internal causes of actions. Folk psychology and natural language are only for gross regularities and coarse classification. They, however, succeed usually in giving information that is quite useful for the understanding of actions.

Now, this picture seems to fit quite well *de dicto* reports, but it would seem doubtful that it does so as well for (what I am calling) *de re* reports. And, of course, it cannot be really made to fit without some modification and filling up.

There is also semantic classification of the agent's internal state involved in such kind of reports. For example, in the report of story 1, the proposition picked up by the embedded sentence as used in the context is the singular proposition that we can represent in the following way:

(5) < <Best-friend, Eleanor, Mary> >

Now, if the report were *de dicto*, this would be part of the content of the belief that the speaker would use in the explanation. In fact he would be relying in more than this. He would be using background information common to speaker and audience—to the effect that Eleanor is in front of (and in full reach, etc.)—for attributing to Mary a belief whose content we could then represent in this way:

(6) < <Best-friend, Eleanor, Mary> > < <In-front-of, Eleanor, Mary> >

Now, this is part of the content of the belief state of Mary that would be used in the explanation of Mary's behavior in story 2. But here in story 1 the speaker is not *attributing* to the hearer a belief with the content of either (6) or (5). Still she is classifying Mary's beliefs with some content or other, the question being which one, and how is it related to the content of the embedded sentence.

I think that what one has to say here is that the content (5) brought out by the sentence, or the content (6) brought out by the sentence and contextual information play only as paradigmatic contents of a class. That is, taking her own intentions into account, the speaker is, roughly, classifying Mary as a person that has in front of her and in full reach her best friend—if her beliefs are correct—, and indicating the relevant classificatory class by "exhibiting"the case itself as a paradigmatic member of it. What allows the speaker to do that is some pragmatic move which relies on the reasonable assumption that, given background knowledge, the "audience" cannot take the report as informing about the specific semantic content of the state. The procedure can still function as an explanation because of the "correlation" between such a type of situation people are in and "warm reaction" behavior, itself also part of shared background knowledge.

Thus, in general, a *de re* report gives the sort of relevant state the agent is in by presenting a representative "picked up" by the speaker having the audience in mind. The implication is that any of the states in the class would lead to the action—given other states the agent is in. *De re* reports are careless about the specific semantic content of the mental states of the agent, being concerned only with presenting such content as a "prototype" of a certain class. But that does not deter them from functioning explanatorily.

I think that somewhat along these lines the postulated primary aim of attitude reports can be reconciled with the claim that there is a basic distinction among them.

I have used only reports with definite descriptions as examples. And although I think that the general position explained here would essentially carry over to other kinds of reports, there are several issues that should be taken care of.

Reports that, instead of definite descriptions, use noun phrases with possessive pronouns as subjects in the embedded sentences, should behave in much the same way. But reports with names or personal pronouns pose an added difficulty. Here, *de re* reports are the "unmarked" case semantically. And then, since the noun phrases contribute semantically just an individual to the content of the embedded sentence as used in the context, and there is no conventionally established "double use," they are good tools for *de re* reporting. What the speaker will then be taken normally as doing in issuing such reports is describing the mental states of the agent only in a "careless" way, unconcerned about their specific content.

There are, however, contrived contexts in which one can use such kind of report for describing with relative specificity the state. Then, even if semantically the key words—names, pronouns—function as having the same content they have when they are not part of such a report, they may have, so to speak, a overimposed pragmatic function. Using shared background knowledge and still other kinds of pragmatic resources, the speaker may succeed in communicating to the audience specific information about the semantic content of the state reported.

I think of the situation as quite similar to cases of potential conflict between semantic and pragmatic factors in the interpretation of definite descriptions. There too we can distinguish between the object (semantically) referred to by the description as referentially used, and what is called the speaker reference, the object which the speaker succeeds in "presenting" to the audience, using pragmatic cues—as the one the utterance is about.[18] In the present case we would have the contrast between the objects the state would be said to be about according to conventional semantic information, and the objects or properties an audience could take the state to be about by using cues from the wording of the report, background information, and still others.

Let me give a final example. Suppose A explains to B the known facts about the Pierre of Kripke's puzzle, and that, on the basis of it, he issues the following complex report:

(7) Pierre believes that London is pretty and he believes also that London is not pretty.

What should be said of such a report? I think that without a context not too far from an explanatory situation, there is not much to feed our intuitions. But let us force the case, just for the sake of illustration. The background—Pierre's story—would make B take the report as attributing to Pierre a state that was being described by its specific content. If so, taking the semantics of the embedded sentences into account, A seems to be *attributing* to Pierre a state semantically characterized by two singular propositions, one of them being the contradictory of the other. B then could ask to A if he is really attributing to Pierre a contradictory belief. If A now answers "yes,"B may disagree about the attribution, quite sceptical about the story justifying it at all. He could, for example, think that the story justifies only—if at all—attributing to Pierre beliefs with singular contents regarding *parts* of London.

If A, however, answers "no,"then B will probably reprimand A by saying: "I knew you did not mean that, but then you should have not worded the report so; "Pierre believes that a city names "Londres" is beautiful, and he believes also that a city names "London" is not beautiful" or something of the sort would have been much more appropriate. The new report—or the "something of the sort"—would make clear that only the attribution of general propositions would then be justified.

Now, as already said, B could as well think, and tell A, that it would also be justified to attribute to Pierre beliefs whose content would be a singular proposition (about, say, particular places or parts of London), but such attributions would not be made correctly with (7). So, it is not that the agent has no "singular thoughts"; it is only that our report does not report about them.

Much work remains to be done in making more precise proposals, but I hope I am not much off the mark when I say that only by putting things in a pragmatic and explanatory context will we find our way out the labyrinth of attitude reports.

NOTES

1. This work is part of the research project PB87-0834-C03-01 supported by the DGICYT.

2. I will not say here anything original. The perspective is the work of philosophers such as Davidson, Dretske, Dennett, and Stalnaker, among others (see especially F. Dretske, *Explaining Behavior*, 1988). Moreover, much of whatever clarity I may have about it at the present time comes from the (unfortunately not yet published) writings of my colleague Manuel Garcia-Carpintero, and also from conversations with him. My exposition will be, as I have already warned, both crude and dogmatic, because, on the one hand, I cannot dwell much in the subject, and, on the other, it involves several of the major philosophical issues relevant for cognitive science that are much discussed in recent years. But I cannot avoid it entirely, since without it as the "background" or "framework" of my discussion, my position would hardly intelligible.

3. As explained—although with a different terminology—by D. Marr in *Vision*, ch. 1. New York: W. H. Freeman, 1982.

4. The first and main thesis about names was established by S. Kripke in *Naming and Necessity*. For the relevance of denominative practices, and a precision of the notion, see G. Evans, *The Varieties of Reference*, ch. 11. For the role of the speaker in determining the referent, see J. Barwise and J. Perry, *Situations and Attitudes*.

5. The distinction was first introduced by K. Donnellan in "Reference and Definite Descriptions," *Philosophical Review*, 75 (1966). As originally introduced the referent of the description in a referential use had not to satisfy the descriptive conditions of the description. I join the majority in not accepting this. A speaker can succeed in pragmatically indicating who or what is she talking about without the object being the object referred to, as would be noticed by possible (mild) protests by an audience (e.g., in replying with "I know who you men, but he is not P," or something of the sort, to a speaker that has said "The P, ...," when the person meant by the speaker does not have property P). In *Situations and Attitudes* Barwise and Perry present a theory of descriptions that

incorporates the distinction without that shortcoming. Their theory allows them to draw a difference among two kinds of attributive uses, what they call respectively the "outer" and the "inner" attributive uses. It seems, then, that the rough distinction of the text between uses that "get us" to a singular proposition vs. uses that "get us" to a general proposition would correspond to their distinction between referential and inner attributive uses.

6. The distinction was first argued in connection with demonstratives by D. Kaplan in *Demonstratives* (unpublished, 1977). J. Perry argued it for personal pronouns in "Frege on Demonstratives," *Philosophical Review, 86* (1977), and "The Problem of the Essential Indexical," *Nous*, 13 (1979).

7. Relevant work about this subject has been made by J. Perry, "The Problem ...," "Perception, Action, and the Structure of Believing," in Grandy & Warner (eds.) *Philosophical Grounds of Rationality*, 1985, G. Evans *The Varieties of Reference*, ch. 6, and C. Peacocke, *Sense and Content*, 1983, ch. 5 and 6, and *Thoughts: An Essay on Content*, 1986, ch. 2.

8. Besides its semantic content, a mental state is differentiated from others by its *kind*: it can be a case of belief, desire, perception, fear, etc. The fact that different states belong to different kinds is traditionally described by saying that the subject can have different *attitudes* towards different contents or even *towards* the same content (as *wishing* to eat a certain cake vs. *fearing* to do it, or *believing* one is to do it.) This accounts for the name. If, moreover, the content is called "proposition," we get the Russellian term, "propositional attitudes," attitudes "towards" propositions. In the second section, there has been some brief talk on the source for differentiating two great kinds of states, doxastic and conative states. That line of thought should be pursued if one wants to explain what is really at stake when the traditional terminology—"attitude towards ..."—is used.

9. Thus we are clearly taking from the outset a realist position on the semantics of attitudes reports.

10. "The Prince and the Phone Booth: Reporting Puzzling Beliefs," Report No. CSLI-88-128. Stanford, CA: Center for the Study of Language and Information, Stanford University.

11. The hypothesis that semantic contents can have such unarticulated constituents has currently wide currency in the work of Perry and his co-workers. See, e.g., "Thought without Representation." *Supplementary Proceedings of the Aristotelian Society*, 60 (1986).

12. S. Kripke, "A Puzzle about Belief," in A. Margalit (ed.), *Meaning and Use* (Dordrecht: Reidel, 1979), pp. 248–249.

13. "The Prince and the Phone Booth," pp. 21–22.

14. The conceptual analysis that leads to maintain that mental states are "constitutively" semantic and what makes us explain action "semantically" must have a common source.

15. Let us assume, for the sake of the example, I am really a competent speaker of English.

16. Note however that this test is not used to define the two kinds of reports, but is only meant to supply some criteria for helping to decide to which kind does a particular report belong.

17. Emphasis in this classificatory viewpoint is due to Perry. See his "Language, Mind, and Information," report no. CSLI-85-44, Stanford University, 1985.

18. See footnote 5.

CHAPTER 9

AUNTY'S OWN ARGUMENT FOR THE LANGUAGE OF THOUGHT

Martin Davies
Birkbeck College, University of London
England

> So we've found at least three reasons for preferring the Language of Thought . . . , and three reasons ought to be enough for anybody's Aunty. (Fodor 1987a, p. 153)

I. INTRODUCTION: AUNTY AND THE LANGUAGE OF THOUGHT

Fodor has launched many arguments in Aunty's direction since *The Language of Thought* (1975), and the assault has not ceased. Even since *Psychosemantics* (1987a), there has been Fodor and Pylyshyn's (1988) attack on connectionism and its followers, amongst whom they number—*qua* "New Connectionist Groupie" (Fodor 1987a, p. 139)—Aunty.

There is something ironic about Aunty's position as the target of this bombardment on behalf of the language of thought (LOT) hypothesis. For there is a strand in Aunty's thought that yields the resources for her own argument for the LOT; and it is an argument with a more *a priori* character than many that she has heard. The greater part of this paper (Sections 2–6) sets out Aunty's argument. That central core is preceded (Section 1) by a discussion of some reservations about the LOT that belong with a different strand in Aunty's thought. The final part of the paper (Sections 7 and 8) is concerned with a further consequence of Aunty's argument: that there is a *prima facie* tension between our commonsense conception of ourselves as thinkers and the connectionist programme for modelling cognitive processes.

Of course, it matters a great deal just what we take Aunty's philosophical background to be. Fodor's Aunty—who, as Dennett (1991) points out, is not to be confused with Fodor's Granny—"speaks with the voice of the Establishment" (Fodor 1987a, p. 135). Myself, I think of her as a rather Oxonian Aunty (an Oxford Aunt, one might say): she has residual Wittgensteinian tendencies, but is fundamentally a neo-Fregean.

As a neo-Fregean Aunty, she holds it as a deep truth that thoughts are structured: thoughts have concepts as constituents, and these constituents can be recombined into other thoughts. Just what this deep truth amounts to has to be made clear; but it is in this area that we shall find the premises for Aunty's argument.

On the other hand, as a residually Wittgensteinian Aunty, she has strong intuitive reservations about the idea of a LOT. At the heart of these reservations lies the idea that the LOT hypothesis is potentially regressive.

II. THE THREAT OF REGRESS

The worry about regress arises in the following way. A sentence is *inter alia* a syntactic object. When a sentence of a public language is presented to someone who understands the language, he is able to assign a meaning to the sentence, and thus to take the sentence as a semantic object. So—Aunty wants to know—if thinking involves tokens of sentences in the language of thought (LOT), for whom are these sentences objects? To whom are these inner sentences presented *inter alia* as syntactic items standing in need of interpretation?

To answer that the LOT sentences are presented in this way to the thinker himself seems quite unsatisfactory, for it involves a regress of languages: a metalanguage of thought in which to think about the sentences in one's language of thought, a metametalanguage, and so on. On the other hand, the answer that the sentences of a thinker's LOT are presented as syntactic items to a little man who reads and understands what is written on some inner blackboard also seems hopelessly regressive. Since understanding itself involves thinking, the little man would need to have his own inner blackboard, and a further, smaller, little man to read what is inscribed upon it.

It seems to Aunty, once this worry takes hold, that the thesis that thinking requires an inner language of thought is about as philosophically disreputable as the thesis that picking a red flower requires an inner collection of colour cards (Wittgenstein, *Blue Book* (1969), p. 3).

Avoiding the Threat

But the worry should not be allowed to take hold. The LOT hypothesis is not regressive.

In order to see this, we need to know what a minimal version of the hypothesis says. The statement that thoughts have syntactic properties is, by itself, unclear. We must distinguish between two senses of the word "thought," and say which properties count as syntactic.

The word "thought" can be used to apply to thought contents or to thought states. If thought contents are the focus of attention, then it is easy to trivialise the LOT hypothesis. For someone might hold that thought contents are canonically specified in a public language like English, and that they simply inherit a syntactic articulation from their linguistic specification. But the LOT hypothesis is not a claim about the structure of our public language descriptions of thinkers. It concerns, not our attributions of mental states, but the structure of the states thus attributed.

If we were primarily concerned with thought attributions in a public language, then the idea of syntactic properties would be unproblematic. But what does it mean to say that certain states of a thinking person have syntactic properties?

From Fodor's work (1987a, pp. 16-21), we can extract three conditions upon syntactic properties. First, a syntactic property is a (higher order) physical property. Second, syntax is systematically related to semantics. Third, a syntactic property is a determinant of causal role or causal powers.

It would be entirely fair to complain that these three conditions are not utterly transparent. For example, Fodor says that shape is the right sort of property to be a syntactic property; and of course, shape is an intrinsic property. So, we might ask whether the first condition is intended to require that syntactic properties of a state are intrinsic, rather than relational; whether this requirement is, rather, an intended consequence of the third condition; or whether the definition of syntax is not itself intended to rule out relational properties. (For a discussion of the notions of syntactic and formal properties, see Devitt 1989.) Likewise, we might ask whether causal powers include both active and passive powers—whether causal role includes both upstream and downstream role.

Let us settle this latter question in favour of just active powers, or downstream causal role, and—ignoring other unclarities—content ourselves with a minimal notion of a syntactic property as a *physical* property that is both *systematically related to semantics* and a *determinant of causal consequences*.

Aunty's argument for the LOT concerns certain states of thinkers. They are states that have semantic properties, and are the inputs to various processors or mechanisms. The conclusion of the argument is that these states have properties that are correlated with their semantic properties and engage those mechanisms. According to the minimal notion of syntax, these properties count as syntactic.

It is easy to see that there is nothing potentially regressive in that conclusion. For the states that have syntactic properties are not presented to anyone—not to the thinker and not to a little inner man—as standing in need of interpretation. The processors which the states engage are not in the position of someone who is presented with a syntactic item but who does not know what to do with it until he knows what it means.

On this point Fodor (1987b, p. 67) says:

> [the formulas of LOT]—unlike those of German—can have their effects on the course of thoughts without having to be understood. This is precisely because—according to the computational story—the psychological effects of Mentalese [LOT] formulas are mediated by their syntactic/intrinsic properties (rather than their semantic/relational properties). This is the trick that computational psychologists use to get the goodness out of postulating a language of thought without having the traditional problem of a regress of languages and interpreters.

It is possible that someone might complain that, precisely because the LOT hypothesis does not give rise to a regress, the terms "syntax" and "language" are misleading since they suggest the case of public language.

There is, perhaps, a hint of this complaint in Barwise's response (1987, p. 83):

> I now realise that for Fodor, the features of language that make "language of thought" an appropriate metaphor have to do with combinatorial-structural properties, whereas what Perry and I have reacted to in the use of this metaphor is the idea that these "expressions" are things that have to be "read" or otherwise "understood" the way expressions of a language used for communication between agents have to be in order to have significance.

To the extent that this complaint is justified, the conclusion of Aunty's argument may not strictly deserve to be called the truth of the *language* of thought hypothesis. But, given that we are explicit about which properties we count as syntactic, and given that it is agreed that the minimal LOT hypothesis is not regressive, there is little here to fight over. Nor is there any need for terminological confusion, provided that we distinguish the LOT from public languages.

The Language of Thought and the Theory of Content

Someone might suggest that this oversimplifies the task of avoiding regress while adopting the LOT hypothesis. For, someone might say, if in the case of the LOT we sever the connection between language and the use of language, then we prevent the LOT hypothesis from making what had seemed to be its distinctive contribution to a philosophical account of mental states.

It can be agreed on all sides that one major task for the philosophy of mind is to give an account of what it is for a mental state to have a particular semantic content. What is it for a belief, say, to be the belief that there are at least three active volcanoes on the far side of the moon, rather than the belief that penguins waddle? The LOT hypothesis seems to promise assistance with this task. For there has been some philosophical progress—in Grice's programme, for example—towards saying what it is for a sentence of a language to have one meaning rather than another. So, if—as the LOT hypothesis says—mental states such as beliefs are relations to inner sentences, then an account of meaning for sentences can be transposed into an account of semantic content for mental states. That would be a significant advance. But the advance is barred if we insist upon a divide between public languages and the LOT. Thus the suggestion.

There are two things that are right about this suggestion. The first is that, if we wish to press a constitutive or analytic account of sentence meaning into service as a theory of mental content, then we cannot also place a chasm between public languages and the LOT. The second is that if we apply an account of sentence meaning in terms of, say, speakers' intentions and beliefs to sentences of the LOT, then the consequence is a regress. If what it is for a sentence of the LOT to have a particular meaning is analysed in terms of intentions and beliefs, then utterances in the LOT must be perpetrated by little men who have intentions to produce beliefs in other little men. And those intentions and beliefs are a matter of relations to sentences in the little men's LOT. And so on, without end.

This might give rise to the thought that the LOT hypothesis has as a consequence that Grice's programme for analysing meaning is circular. But that is not the lesson to draw. Rather, we should indeed distinguish public language from the LOT; and we may, if we are so inclined, pursue Grice's programme with respect to public language meaning (Schiffer 1988, pp. xix-xx). For the LOT hypothesis is not offered as a putative solution to the problem of mental content.

The LOT hypothesis does constrain both the form of the problem and the form of the solution. Since the LOT hypothesis is a form of intentional realism (Fodor 1987a, p. 135), the problem of mental content that it poses is that of providing an account of "x is a belief that p" where "x" ranges over intentional states, rather than merely of "x believes that p" where "x" ranges over intentional systems. And since the LOT hypothesis is not to be regressive, the account should not be cast in terms of the production of these states by communicating agents.

Within those constraints there are many candidate accounts of mental content, including accounts in terms of causal covariance, teleology, and functional role. But the LOT hypothesis is not itself a such candidate.

With all reservations about regress now set aside, we can turn to the structure of Aunty's argument itself. It comes in two main stages. The first stage (Sections 2 and 3) is an argument for a conditional claim: If a cognitive process is systematic—in a sense to be defined—then the inputs to that process have syntactic structure—in the minimal sense just characterised. The second stage (Sections 4 and 5) is an argument for the claim that being a thinker—a believer, a deployer of concepts—involves the systematicity of certain inferential transitions amongst thoughts. Consequently, the inputs to those transitions—thoughts—have syntactic structure; that is, there is a language of thought. (Section 6 deals with two objections to the second stage.)

III. FIRST STAGE: SYSTEMATIC COGNITIVE PROCESSES

The first stage of Aunty's argument involves a conditional claim: If a cognitive process is systematic, then the inputs to that process have syntactic structure. Some of the terms in this claim need a little explanation.

For present purposes, a *cognitive* process is one whose input states have a semantic description, and whose output states have either a semantic description or an action description. A cognitive process thus moves from information to information, or from information to action.

The idea of a *systematic* process that is used in the argument is an essentially relative one: a process is *systematic* relative to a pattern in its input-output relation. Suppose that a generalisation G describes a pattern to be found in the input-output relation of some physical system. If we consider several input-output pairs that exhibit the common pattern, then we can ask whether the several input-output transitions have a common causal explanation corresponding to the common pattern that they instantiate. If there is a common causal explanation, then we can say that the process leading from those input states to output states is causally systematic relative to the pattern described by G.

The Drinks Machine

Consider the following very simple example. There is a machine that produces coffee or tea with or without milk; the output states of the machine are states of delivering drinks of one of four kinds. The input states of the machine are states of having a token of one of four kinds in its slot.

The four kinds of token are these. There are square red tokens, square blue tokens, round red tokens, and round blue tokens. If a square red token is in the slot, then the machine delivers coffee with milk. If there is a square blue token in the slot, then the machine delivers coffee without milk. If there is a round red token in the slot, then tea with milk is delivered. And if there is a round blue token in the slot, then tea without milk is the result.

Under these descriptions of input and output states, there is a clear pattern to be discerned in the input-output relation for the drinks machine. Whatever the colour of the token (whether it is red or blue), if it is square then coffee is delivered, whereas if it is round then tea is delivered. And whatever the shape of the token (whether it is square or round), if it is red then a drink with milk is delivered, while if it is blue then a drink without milk is delivered.

We can ask whether the process that mediates between input and output states is causally systematic relative to the pattern described by each of the four little generalisations about the machine's input-output relation.

Thus, for example, is there a common explanation for the delivery of coffee consequent upon the input state of having a red square in the slot and the input state of having a blue square in the slot? Is there a common processor or mechanism responsible for mediating these two transitions? Likewise, is there a common explanation for the inclusion of milk in the drink delivered when a square red token or a round red token is inserted in the slot?

The answers to these questions are not determined by the facts about the input-output relation, but by facts about the internal architecture of the drinks machine. One possible configuration would have, within the machine, four autonomous and totally dedicated drink producing devices: one activated by each of the four possible input states. Another possible internal configuration would have three component devices. First, there would be a device that is activated by either a square red token or a square blue token in the slot, and produces coffee. A second device would produce tea if there is a round red or round blue token in the slot. And a third device would add milk to the drink produced if there is a round or square red token in the slot but refrain from adding milk if there is a round or square blue token in the slot.

These two configurations would license different answers to the question about causal systematicity. The operation of a drinks machine with the first configuration is not causally systematic relative to the input-output patterns that we discerned, whereas the operation of a machine with the second configuration is systematic.

As this example illustrates, if we think of a physical system as containing various subsystems or mechanisms, then the requirement for causal systematicity relative to the pattern described by G is that there should be a mechanism whose presence in the system explains all the input-output transitions that conform to the pattern described by G. It is not sufficient that this common mechanism should merely figure as a component somewhere along the way in the several transitions. Rather, the common mechanism should actually mediate between inputs and outputs in accordance with G.

The example also illustrates that conformity to an input-output pattern is in no way sufficient for causal systematicity, as that notion is deployed in Aunty's argument for the LOT. There can be two systems with the same input-output relation, where the processing in one system is causally systematic relative to some pattern in that input-output relation, while the processing in the other system is not systematic relative to that pattern. The distinction here is one of which we make widespread use in our descriptions of complex systems.

The conditional claim to be established in this first stage of the argument concerns systematic cognitive processes. These are processes that are systematic relative to patterns revealed under semantic descriptions of the input and output states (or semantic descriptions of the input states and action descriptions of the output states). The semantic description of the input states is crucial to the conditional claim since, according to the notion of syntax that we are using, there are no syntactic properties without semantic properties. (For other purposes, a different notion of syntax would be appropriate. See, for example, Stich 1983; and, for some problems over the idea of syntax without semantics, Crane 1990.)

For a simple example in which the input states do have semantic descriptions, we can return to the drinks machine.

The Drinks Machine Again

Let us speak in a pretheoretical way, and say that the machine's input state of having a square red token in the slot means that the client wants coffee with milk, the presence of a square blue token in the slot means that the client wants coffee without milk, and so on.

Given these semantic descriptions of the input states and the action descriptions of the output states, we can redescribe the pattern in the input-output relation of the drinks machine. If the input state means that the client wants coffee (whether with or without milk), then the output state is a delivery of coffee. Similarly, if the input state means that the client wants a drink with milk (whether it be tea or coffee), then milk is included in the drink that is delivered. And so on.

As before, we can ask whether the operation of the machine is causally systematic relative to each of the patterns described by these generalisations. One possible internal configuration of the machine will warrant a negative answer; another will warrant an affirmative answer.

The Sentence Interpreter

Consider now the cognitive process of understanding some English sentences. To be more accurate, what is to be considered is the process that begins with a state registering the information that a particular sentence has been uttered, and ends with a state registering information as to what has been said - what message has been conveyed.

If you understand the three sentences, "Martin is tired," "Martin is tall," "Martin is drunk," then in each case you end up knowing that what has been said concerns this person here. We can describe a pattern in the input-output relation: If the input state registers the utterance of a sentence that contains the name "Martin," then the output state means that what was said was about this person here. Likewise, we can describe patterns relating to the messages conveyed by other sentences: "Martin is tired," "Andy is tired," "Frank is tired." If an input state registers the utterance of any of these sentences, then the output state that is produced means that what was said was to the effect that someone (whether it be this person here, or . . .) is tired.

The causal systematicity of this cognitive process requires more than just conformity to these patterns. Systematicity relative to these generalisations requires that, corresponding to each pattern (in extension) there should be a common mechanism whose presence explains the aspect of input-output transitions that is captured in that pattern. Within the physical system that performs the transitions that we are calling sentence interpretation, there should be a component mechanism that is responsible for mediating transitions between input states registering the utterance of sentences containing the name "Martin" and output states that concern this person here. Likewise, there should be a component mechanism that is responsible for mediating the several transitions from input states that concern sentences containing the predicate "is tired." The interpretation of the sentence "Martin is tired" will then be the joint product of those two mechanisms. Causal systematicity thus requires real commonality of process.

Knowledge of Rules

The idea of causal systematicity is also involved in the account that I would give of knowledge of rules (Davies 1986, 1987, 1989, 1990a, 1990b). Where there is causal systematicity relative to a pattern revealed under a semantic description of the input and output states, there the system has knowledge of the rule or generalisation describing that pattern.

The example of the sentence interpreter provides a straightforward case. Knowledge of the rule that sentences containing the name "Martin" convey propositions about this person here does not require the ability to formulate explicitly the thought that is indeed a rule of the language in question. What it does require is causal systematicity relative to the input-output pattern described by that rule.

The conditional claim with which the first stage of Aunty's argument is concerned can then be stated in terms of knowledge of rules: If a cognitive processing system embodies knowledge of a rule, then the input states of the system have syntactic structure. This conditional claim does not say that if a process is causally systematic, and so involves knowledge of a rule, then the process operates in virtue of an explicit syntactic encoding of the rule that is known. For all that the conditional claim says (and for all that Fodor says—see 1985, p. 95; 1987a, p. 25) the standing condition of knowledge of a rule can be realised just as well by the presence of a component processor as by the presence of an explicit representation. The consequent of the conditional claim concerns just the input states of cognitive systems.

IV. FIRST STAGE: FROM SYSTEM TO SYNTAX

We now have some grasp upon both the antecedent and the consequent of the conditional claim. The notion of causal systematicity is a relative one, and the cases that concern us involve systematicity relative to patterns that are revealed when the input states (at least) are given semantic descriptions. The minimal notion of a syntactic property that we are using is also a relative one; in fact, it is doubly relative.

First, what counts as a syntactic property depends upon what semantic properties are present, since a syntactic property must be systematically related to semantics. Second, what counts as a syntactic property of an input state depends upon the actual constitution of the machine of which it is an input state. For a syntactic property must be a determinant of causal powers, and a property to which the operation of one machine is sensitive may be quite irrelevant to the operation of another machine.

Of course, it is one thing to understand a claim, and quite another to have an argument for its truth. We can proceed a good part of the way towards seeing why the conditional claim is true if we return to the first example of the drinks machine, in which the input states are described simply in terms of the shape and colour of the tokens in the slot. For we can observe that causal systematicity of process imposes requirements upon the causal properties of the input states.

Suppose that the operation of the drinks machine is causally systematic. Then there is *inter alia*, as a component of the drinks machine, a common mechanism that operates to mediate the transition from either a square red token or a square blue token in the slot to the delivery of coffee. But then there must be some property shared by—and distinctive of—those input states which is causally adequate to engage that mechanism.

There must also be a causally relevant difference between those two input states, since one state engages the milk introducing mechanism while the other does not. Indeed, in order to engage that mechanism, the input state of having a square red token in the slot must have some causal property in common with the input state of having a round red token in the slot—a property not shared by the other two input states. In short, the input states exhibit patterns of recurrent properties that are determinants of the causal consequences of those states in the context of the drinks machine.

It is, of course, an empirical question what the causally salient properties of the input states are. All that causal systematicity requires is that the operative properties of the input states should correlate with the properties cited in the descriptions of the input-output patterns. It might be that the squareness of tokens in the slot is what engages the coffee mechanism, and that the redness of tokens is what engages the milk mechanism. But it might also be that the square tokens or the red tokens have something else in common, such as a distinctive mass, a distinctive chemical composition, or a distinctive inscription upon them.

The properties required by causal systematicity in the first example of the drinks machine do not yet qualify as syntactic properties, since no semantic properties have been introduced for them to be correlated with. But we can take the final step towards seeing why the conditional claim is true if we now consider the second example involving the drinks machine. In that example, the input-output pattern is revealed under a semantic description of the input states.

So, suppose that the operation of the drinks machine is causally systematic relative to the patterns that are revealed under semantic descriptions of the input states. This is to say that the machine has knowledge of rules such as: Deliver coffee, given that the client wants coffee.

This causal systematicity imposes a requirement that the input states that mean that the client wants coffee (whether with or without milk) should have a causal property in common, in virtue of which those input states engage the coffee producing mechanism. Similarly, the input states that mean that the client wants a drink with milk should have in common a property that engages the milk introducing mechanism. There is no particular requirement as to what these properties should be; they might or might not be the squareness and redness, respectively, of the token in the slot. But, over the range of input states, they have to correlate with meaning that the client wants coffee and meaning that the client wants a drink with milk respectively. Exactly similar considerations apply to the example of the sentence interpreter.

In short, what we see is that causal systematicity relative to semantic input-output patterns (or equivalently, knowledge of rules) requires that the input states of the machine should have properties that are correlated with their semantic properties, and are determinants of the causal consequences of those states given the internal constitution of the machine. Since these will surely be physical properties, they will meet all three conditions on syntactic properties.

There are two points to notice about the conditional claim. The first point is that the complexity of the syntax of input states may be very modest indeed. For example, in the case of the drinks machine, the formal language of its input states has just four primitive symbols and one binary operation; and the operation does not even distinguish the order of constituents. The second point is that not every aspect of semantic content is required to be articulated syntactically (*cf.* Perry 1986). The argument for the conditional claim requires no syntactic property corresponding to an aspect of semantic content that is constant over all input states of the machine (such as that all input states mean that the client wants something). The drinks machine is dedicated to the wants of the client; and task dedication permits syntactic inarticulateness.

So much, then, for the first stage of Aunty's argument for the LOT hypothesis. Its plausibility depends upon two things. A minimal notion of syntactic property is used in the consequent of the conditional; and the notion of causal systematicity that is used in the antecedent requires much more than just that a pattern (in extension) should be exhibited by an input-output relation. In short, the truth of the conditional claim is secured by having a relatively strong antecedent and a relatively weak consequent. The price of this strategy is, of course, that it increases the burden upon the second stage of the argument.

The task of the second stage of Aunty's argument is to uncover, in the commonsense scheme of thought, concepts, and inference, a commitment to causal systematicity of cognitive processes. That second stage depends upon Aunty's neo-Fregean conception of thoughts.

V. SECOND STAGE: THE STRUCTURE OF THOUGHT

Thoughts are states with semantic content, and consequently with truth conditions. But thoughts are not the only semantically evaluable states in the world; nor are they the only psychological states that have semantic content. Having content is a feature that is shared by thoughts, by certain patterns of sound waves and marks on paper, by states of the visual system of humans and other animals, by patterns of tree rings, and by states of room thermostats.

This is not to say that thoughts have content in just the same way as these other states. Someone might hope that there can be a unified theory of all these contents, but there are reasons to think that we need to distinguish between information content—for which some causal—cum—teleological theory might be right—and mental content (the content of propositional attitudes)—for which such a theory is inadequate.

If this is right, then, amongst the psychological states of a person, we need to distinguish between states with mental content and states with mere informational content—between propositional attitude states and *subdoxastic* states (Stich 1978). And a substantial project in the philosophy of psychology is to give a principled account of this distinction.

One way to commence on that project is to focus upon the fact that thoughts—and attitude states in general—are states whose semantic content is *conceptualised* content. A person who is in such a state *ipso facto* deploys the constituent concepts of the content of that state (*cf.* Davies 1989). This is not so for states of early visual processing, for example.

The "ipso facto" is important; for certainly a person may have a thought about the content of a state of visual processing, and so conceptualise the content of that state. Similarly, a theoretical linguist may have a thought about the information content of some state of the language system. But having those thoughts is not essential to being in the respective states.

The content of thoughts is conceptualised content. To entertain a thought, to hold a belief, or to frame an hypothesis, involves deployment of concepts. Thus, no one can entertain a thought with a particular content without grasping the constituent concepts of that content. Furthermore, for a thinker to have the concept of being F, the thinker must *know what it is for* an object to be F—that is, know what it is for an arbitrary object to be F. (This epistemic requirement for possession of the concept of being F has an analogue for thoughts about particular objects; namely, that the thinker should *know which* object is in question. Gareth Evans (1982, p. 65) calls this *Russell's Principle.* As Evans points out (*ibid,* pp. 76-9), these requirements involve rejection of the *Photograph Model* of mental representation.)

Putting these ideas together, we arrive at an important neo-Fregean consequence. To entertain the thought that object *a* is F a thinker must have the concept of being F. If a thinker has that concept, then the thinker knows what it is for an arbitrary object to be F. So, if a thinker thinks that *a* is F and is able to think about the object *b*, then the thinker is able to entertain the thought—to frame the hypothesis—that *b* is F.

This consequence is, in effect, what Evans (*ibid,* p. 104) calls the *Generality Constraint;* and it has as an immediate consequence (perhaps not properly distinguishable from the Generality Constraint itself) a *closure condition* on the domain of thought contents available to a thinker.

If a thinker can be credited with the thought that object *a* is F and the thought that object *b* is G, then that thinker has the conceptual resources for also entertaining the thought that *a* is G and the thought that *b* is F. Similarly, if a thinker can be credited with the thought that *a* is R to *b*, then that thinker has the conceptual resources for also entertaining the thought that *b* is R to *a*. The domain of thought contents available to a thinker is closed under recombination of conceptual constituents.

Thoughts are states with semantic content, and these contents are of a special kind that is subject to the Generality Constraint and thus to the closure condition. These are two important neo-Fregean claims. But they leave us some way from causally systematic processes; and they do not themselves permit a direct argument for the LOT.

Semantic Content and the Closure Condition

Consider first the claim that thoughts are states with semantic content. Certainly there is no argument from the mere fact of semantic content to the LOT hypothesis.

Suppose that a creature evolves in an environment where the prime predatory danger typically arises after a hawk dives on a beetle. Suppose that the creature develops a detector for just this scenario: a hawk-diving-on-beetle detector. Suppose that the detector operates by being sensitive to overall aspects of the threatening scenario; and not by being composed *inter alia* from a hawk detector and a beetle detector. It is immensely plausible that, for a causal-cum-teleological notion of information content, there is a state of this creature with the semantic content that a hawk is diving on a beetle. But there is no reason at all to suppose that the state has a syntactic constituent structure.

Consider second the claim that thought contents are subject to the closure condition. There is no utterly compelling argument from the closure condition on semantic contents to the LOT hypothesis.

One manifestation of the closure property is that if a system has a state with the content that *a* is R to *b*, then it also has a state with the content that *b* is R to *a*. But the availability of states with these contents does not require a syntactically structured vehicle for semantic content.

Thus suppose that, for whatever reason, our creature with a hawk-diving-on-beetle detector also develops a second detector for a second threatening scenario. Suppose that danger is often just around the corner when a beetle dives on a hawk; and that the creature consequently develops a beetle-diving-on-hawk detector. Let it be that these two proprietary detectors—with their downstream processors that produce appropriate evasive behaviour—are causally autonomous from each other; indeed, we could think of each one as a module within the creature's total information processing system.

This example has been set up so that there is no common syntactic constituent in these two information registering states. There is no syntactic symbol meaning *hawk* that is implicated in the two states, for example. For the two states to have a syntactic constituent in common would require there to be some common property of the two states which is systematically related both to the semantic content of the two states and to the causal consequences of the two states. There is, to be sure, a common property of the two states which is related to their semantic contents; namely, a complex relational property having to do with the causal antecedents of the states. But this property is not directly implicated in the production of the causal consequences of the states.

This example of the two detectors is, of course, a mere toy. The idea behind the example can be extended to some other, more complex, toys, such as the sensori-motor coordination system in Paul Churchland's (1986) crab. But, with or without further examples, the principle is clear. *De facto* compliance with the closure condition does not inevitably require syntax.

It might be responded here that it is significant that all the examples that are supposed to illustrate this principle share the feature of being only toy examples. The claim might be made that once we attempt to meet the closure condition for a suitably rich set of semantic contents, without overreaching the available computational resources, we shall inevitably have to make use of some syntactic articulation. For, it might be asked, *how else* could we do it?

This is quite a potent challenge. Aunty is not obliged to deny that —properly developed—it can establish a strong plausibility consideration in favour of the LOT hypothesis. Nevertheless, a "How else?" challenge is always open to the risk that someone will respond: "Like so." This is what connectionists, for example, do.

The LOT hypothesis is supposed to play a pivotal role in an argument for a *prima facie* tension between our conception of ourselves as thinkers and the connectionist programme. Consequently, the argument from our commonsense conception to the LOT must not appear to beg the question against connectionism. What is needed, and what Aunty offers, is an argument that is more direct and a *prioristic* than any "How else?" challenge.

To the extent that Aunty does not rest content with a "How else?" challenge, she also does not rely upon an argument by analogy that proceeds as follows.

The semantic contents of sentences of a natural language meet a closure condition. If, for example, there is a sentence that means that *a* is R to *b*, then there is also a sentence that means that *b* is R to *a*. The sentences of a natural language meet this closure condition by having a syntactic constituent structure. The contents of thoughts meet a similar closure condition. Therefore, by analogy, thoughts have syntactic constituent structure, too.

The analogy between the meanings of natural language sentences and the contents of thoughts is not perfect. After all, it is possible to have mere phrasebook mastery of a (fragment of) a language; whereas it is not possible to have phrasebook mastery of thoughts (*cf.* Evans 1982, p. 102). But, once again, it is not necessary for Aunty to hold that the argument is totally without merit. It is simply that the argument by analogy cannot serve our dialectical purpose; the most that it can achieve is to establish a plausibility consideration, pending the investigation of alternative vehicles of semantic content.

VI. SECOND STAGE: CONCEPTS AND INFERENCE

The conclusion of Aunty's argument cannot be reached directly from the claims that thoughts have semantic content, and that those contents are subject to the closure condition. But, fortunately for Aunty, those two claims do not exhaust the significance of the neo-Fregean idea of conceptualised content.

It is a feature of the thought that *a* is F that entertaining that thought involves mastery of the constituent concept of being F; a piece of concept mastery that can be employed in further thoughts about other objects. So, it is not merely the case that if a thinker can think that *a* is F and think that *b* is G, then he can also think that *b* is F. It is not merely that there is one state of the thinker with the content that *a* is F and another state with the content that *b* is F. Rather, entertaining the thought that *a* is F and entertaining the thought that *b* is F involve the deployment of a common piece of concept mastery—mastery of the concept of being F—and a common piece of knowledge—knowledge what it is for something to be F.

This is part of what is involved in the idea of conceptualised content; but it is not captured by the closure condition, since that condition could be met by the occurrence of states that are quite autonomous. The closure condition would be satisfied provided that, whenever there are states with the contents that *a* is F and that *b* is G, there are also states—even states that are intrinsically quite unrelated—with the contents that *a* is G and that *b* is F.

If we take the claim about common pieces of concept mastery, and combine it with the familiar picture of thoughts related in an inferential web, then we can derive a consequence that is highly promising given the purposes of our argument. Indeed, this consequence is explicit in Evans (1981, p. 132):

> To have a belief requires one to appreciate its location in a network of beliefs. . . . To think of beliefs in this way forces us to think of them as structured states; the subject's appreciation of the inferential potential of one belief (e.g. the belief that a is F) at least partly depending upon the same general capacity as his appreciation of the inferential potential of

> others (e.g. the belief that b is F). . . . Possession of this general capacity is often spoken of as mastery of a concept.

A thinker who has the thought that *a* is F appreciates that from this thought it follows that *a* is H, say; and he also appreciates that from the thought that *b* is F it follows that *b* is H. But that is not all. It is not just that there is an input-output pattern in the inferences that the thinker is disposed to make. The two inferences are manifestations of a common underlying capacity; namely, mastery of the concept of being F.

As Evans himself makes clear, the notion of a capacity or disposition is not to be understood in terms of the bare truth of conditional statements, but rather in a "full-blooded" way (1981, p. 329). The idea of a common capacity being manifested in the two inferences should be unpacked in terms of a common explanation, adverting to a common state (1982, p. 102). In short, there is causal systematicity relative to the input-output pattern in a thinker's inferential practice.

Here is a simple example. A thinker who has the thought that Bruce is a bachelor appreciates that from this thought it follows that Bruce is unmarried; he also appreciates that from the thought that Nigel is a bachelor it follows that Nigel is unmarried. The thinker appreciates the inferential potential of the two thoughts; and this depends in each case on the same general capacity, namely, mastery of the concept of being a bachelor.

In order to have either the thought that Bruce is a bachelor or the thought that Nigel is a bachelor, the thinker must grasp the concept of being a bachelor. This is a matter of knowing what it is for an object to be a bachelor; of knowing *inter alia* that to be a bachelor requires being unmarried. This single piece of knowledge—that for an arbitrary object to be a bachelor, that object must be unmarried—is implicated in both the inferential transitions that the thinker is disposed to make.

All this is just what is needed for the second stage of Aunty's argument. It is part of the neo-Fregean conception of a thinker that, in the arena of thought, there is a genuine causal systematicity of inferential transitions.

VII. TWO OBJECTIONS TO THE SECOND STAGE

Aunty's argument is essentially complete. But we should pause to consider two objections that might be raised against the second stage of the argument. One concerns the appeal to Evans's work; the other begins from the very simple example of thoughts about being a bachelor.

Evans on The Language of Thought

It might be objected that there is something infelicitous about Aunty's reliance upon Evans. For (as critics of Aunty's argument sometimes point out) Evans himself says (1982, pp. 100-101):

> It seems to me that there must be a sense in which thoughts are structured. . . . This might seem to lead immediately to the idea of a language of thought, . . . However, I certainly do not wish to be committed to the idea that having thoughts involves the subject's using, manipulating, or apprehending *symbols*—which would be entities with non-semantic as well as semantic properties, . . . I should prefer to explain the sense in which thoughts are structured, not in terms of their being composed of several distinct *elements*, but in terms of their being a complex of the exercise of several distinct conceptual *abilities*.

In this passage (just before the introduction of the Generality Constraint) Evans rejects outright a certain conception of the LOT hypothesis—according to which it involves *the subject's* using symbols—and he denies that the idea of thoughts as structured leads *directly* to the LOT hypothesis.

But, on these two points, Aunty is in total agreement. First, it is certainly no part of the LOT hypothesis for which Aunty argues that the conscious, thinking subject is presented with thoughts as entities with non-semantic properties. Indeed, that would arguably be regressive. The LOT hypothesis concerns the scientific psychological underpinnings of a subject's conscious mental life.

Second, Aunty's argument does not move immediately to the LOT hypothesis from the idea of thoughts as structured. Rather, the argument follows Evans in moving first to the notion of the exercise of common capacities, and in construing capacities in a full-blooded way. The step from there to the LOT involves the conditional claim linking systematicity of process with syntactic structure in input states (established in the first stage). That step of the argument is not, apparently, anticipated by Evans; but nor is it considered and rejected by him.

In short, then, there is less to the difference between Aunty and Evans than the quoted passage suggests.

Concept Mastery and Primitively Compelling Inferences

There are many things that a thinker might conclude about Bruce, given that Bruce is a bachelor, which he would not conclude about Nigel, given the thought that Nigel is a bachelor. A thinker might reasonably conclude that Bruce drinks a lot of Foster's Lager, while Nigel drinks a lot of Spanish champagne, for example. And where a thinker does draw similar conclusions about Bruce and Nigel, there is not, in general, any guarantee that the inferential transitions are products of a common capacity.

Consequently, the plausibility of the idea of causally systematic inferential transitions might seem to be an artifact of the definability of being a bachelor. So it might be objected that, for almost any concept other than the concept of being a bachelor, it would be implausible to insist that there is causal systematicity of inferential transitions.

This second objection might be coupled with the idea that, where mastery of a concept does not consist in knowing a definition, objects falling under the concept exhibit only a family resemblance. And that might be thought to undercut even further the idea of a common capacity being exercised in inferences concerning different objects.

However, we can respond to this objection by noting that mastery of a concept may be constituted by commitment to a set of inferential principles, without those principles amounting to a statement of necessary and sufficient conditions for application of the concept. An alternative development of the argument for causal systematicity illustrates this point.

In recent work, Christopher Peacocke (1986, 1989a, 1990) has been articulating a theory of concept mastery. In *Thoughts* (1986), the idea is expressed in terms of the canonical grounds and the canonical commitments of certain classes of contents. In "What are concepts?" (1989a), the idea is of a possession condition for a concept, where this is often a matter of a thinker finding certain patterns of inference primitively compelling. (For a brief account, see Peacocke 1989b.)

Here is an example that Peacocke (1990) develops. Part of what is involved in mastery of the concept *plus* is finding this inferential transition (T) primitively compelling:

> 18 + 64 is a certain number n;
> so
> 18 + (the successor of 64, *viz.* 65) is the successor of n.

Likewise, the master of *plus* finds this transition (T′) primitively compelling:

> 11 + 23 is a certain number m;
> so
> 11 + (the successor of 23, *viz.* 24) is the successor of m.

And, of course, there are indefinitely many other primitively compelling inferential transitions exhibiting the same pattern.

Now, on Peacocke's account, mastery of the concept *plus* involves more than that the thinker should find each of these transitions—or enough of these transitions—primitively compelling. What is required is that the thinker should find the transitions primitively compelling *in virtue of their form.*

This is not intended as the requirement that the thinker should be able to conceptualise or to formulate the form or pattern of inference (R):

> From: $m + k = n$;
> Infer: $m + S(k) = S(n)$.

Rather, the account is supposed to relate "grasp of *plus* to the causal influence of a form of transition which is not necessarily conceptualized" (p. 65). The idea is that the form of transition is causally explanatory: it enters the causal explanations of particular transitions being found primitively compelling. And this phenomenon —of causally explanatorily relevant forms or patterns—is one that can be found in humans and in machines.

In this alternative development of the second stage of Aunty's argument, Peacocke's idea of inferences that are found primitively compelling in virtue of their form—of a causally influential form of transition—is glossed in terms of causal systematicity of process.

The proposal is that at least part of what is involved in particular inferences being found primitively compelling in virtue of their form is this. Mirroring the commonality in the inferences that are found primitively compelling—namely, their form—there should be a commonality in the causal processes that explain their being found so.

Given this gloss, and the close connection between causal systematicity and knowledge of rules, we might call the common state that figures in the causal explanations of the various particular inferences—such as (T) and (T′)a state of knowledge of rule (R). As Peacocke says, this knowledge of (R) does not require that (R) should be conceptualised or formulated by the thinker. Nor does it require that (R) should be explicitly represented in the thinker's cognitive machinery.

Provided that the state of knowledge of an inferential rule such as (R) functions to mediate actual transitions in thought—from the premise of (T) to the conclusion of (T), from the premise of (T′) to the conclusion of (T′), and so on—we have here an alternative version of the second stage of Aunty's argument. It is a version that can subsume the simple cases such as the concept bachelor, without suggesting that its application is restricted to concepts that are definable.

Taken together with the conditional claim of the first stage, it requires that the input states of the transition mediator corresponding to an inferential rule should have a syntactic articulation. Thus it provides an alternative completion of Aunty's own argument for the LOT.

VIII. CONCEPTUALISED THOUGHT AND THE CONNECTIONIST PROGRAMME

Given Aunty's argument that our commonsense conception of ourselves involves a commitment to the LOT hypothesis, we can now argue for a *prima facie* tension between that commonsense conception and the connectionist programme for modelling cognitive processes.

This argument turns upon the claim that typical connectionist networks do not exhibit causal systematicity of process, and syntactic structure in input states. Of course, connectionism comes in several varieties, and there are some networks that do have these features; examples can be provided by networks with local representation of all the primitive concepts of some classical task analysis. So let us be more specific. What is to be considered is connectionism with distributed representation. In particular, we focus on networks with microfeatural, dimension shifted, representation in the style of Smolensky (1988).

We can begin with the question whether connectionist networks have syntactically structured input states. Having in mind that syntax is relative to semantics, we should be explicit. The question is whether the input states of a network have syntactic structure relative to the standard or classical semantic description of what the network is doing. Are there properties of connectionist input states, as such, which line up with the primitive concepts used in a classical analysis of the task that the network is performing?

Syntax

If the representation in a connectionist network is distributed rather than local, then activation at an individual input unit cannot be regarded as the tokening of a syntactic element. The reason is simple; for to say that representation is distributed is just to say that individual units are not the vehicles of representation.

Thus, the input states of a network may be representing facts about coffee in various contexts: in cups and jugs, with or without sugar.

But there is not an individual unit that represents the occurrence of coffee. The representation of coffee in a cup is not a matter of activation at a coffee unit and a cup unit. Rather, what represents coffee in a cup is a pattern of activation over many units.

However, this simple fact about distributed representation does not yet show that there is no syntactic description of connectionist input states. Activation at a single unit is just a limiting case of a subpattern of activation; and which units are included in a total pattern of input activation is certainly a determinant of the causal consequences of that state. So—given our minimal notion of syntactic property—a subpattern of activation over several units is the right sort of thing to count as the tokening of a primitive symbol, provided that the subpattern corresponds to a semantic property of the input states in which it occurs.

So, although there is no proprietary coffee unit, might there not be a specific distributed pattern of input activation that means coffee?

There certainly are networks that show subpatterns of input activation of the envisaged kind. In the networks examined by Ramsey, Stich and Garon (1990), the input states that represent various propositions about dogs all share a common subpattern of activation over eight input units, given by the vector <11000011>. Similarly, the activation vector <11001100>, over those same eight units, occurs whenever the proposition concerns cats, while whenever the proposition concerns having fur there is a common subpattern of activation over the remaining eight input units, given by the vector <00001111>. Consequently, the pattern of activation

<1100001100001111>

for the proposition that dogs have fur, for example, can be regarded as the tokening of two primitive symbols, with co-occurrence of subpatterns being the network's way of combining subject and predicate to make a sentence.

Nevertheless, what Smolensky says indicates that he does not see this as the typical case. Concerning the constituent subpatterns of activation that represent coffee in various contexts—coffee with sugar, coffee in a cup, coffee in a jug—Smolensky says (1988, p. 17):

> These constituent subpatterns representing coffee in varying contexts are activity vectors that are not identical, but possess a rich structure of commonalities and differences (a family resemblance, one might say).

(If we focus upon representations of coffee in a cup, coffee in a jar, and coffee on a tree, then we invite the response that, even by classical lights, different concepts are involved: coffee drink, coffee granules, and coffee beans. But the point about contextual variation of microfeatural representation carries over to the cases where that response is no longer plausible.) So, there simply is no strictly common subpattern of activation that can be identified as a syntactic element meaning coffee.

If Smolensky is right then, relative to a semantic description in terms of coffee, cups, jugs, and the like, the input states of a typical connectionist network with distributed representation will not have a syntactic description.

Essentially the same point could be made in terms of a network performing the sentence interpretation task sketched in Section 2. We can imagine that the input states that register which sentence has been uttered (or presented visually) make use of distributed microfeatural representation. And we can suppose further that the way in which the predicates, for example, are pronounced (or written) varies somewhat, depending upon the name with which they are combined. Consequently, it may be that the input representation of the occurrence of the predicate "isdrunk" varies between its occurrences in the contexts "Martin is drunk," "Andy is drunk" and "Frank is drunk." The constituent subpatterns may exhibit family resemblance, rather than identity. In that case, although the objects in the task domain have syntactic structure, the input states of the network will not (relative to their semantic description as representing the names, predicates, and sentences of the task domain).

Systematicity

This fact, that networks of a certain type do not have syntactically structured input states, need not threaten the conditional claim in the first stage of Aunty's argument for the LOT hypothesis. That conditional claim remains intact, so long as the networks under discussion do not exhibit causal systematicity of process (relative to input-output generalisations pitched at the level of semantic description in the terms of a classical task analysis). And, indeed, they do not.

Let us return to our hypothetical example of a network whose input states represent facts about coffee in various contexts. We can imagine that some of the input states mean, variously, that there is coffee in a cup, coffee in a jug, coffee in a glass, coffee with sugar, and some that there is wine in a cup, wine in a jug, wine in a glass, wine with sugar, and so on. Similarly, we can imagine that some of the output states mean that there is warm drink in a cup, warm drink in a jug, warm drink in a glass, warm drink with sugar, and the like.

Now suppose that the network is performing some rudimentary inferential transitions. The input state with semantic content *coffee in a cup* produces the output state with semantic content *warm drink in a cup;* the input state with content *coffee with sugar* produces the output state with content *warm drink with sugar;* and so on. Viewing this from the outside, we can see a pattern in the input-output relation for this network: Whenever the input state means *Y (coffee),* the output state means *Y (warm drink).*

To ask whether the process that is going on in the network is causally systematic relative to that pattern is to ask whether the *coffee* to *warm drink* transitions all have a common explanation; whether there is, as a component of the network, a mechanism that is responsible for all and only those transitions.

In general, the answer to this question is negative. It will not be strictly true that there is a common set of weights on connections that is implicated in all and only the *coffee* to *warm drink* transitions. In terms of knowledge of rules, we can say that it will not be correct to describe the network as having knowledge of the rule:

Given: there is coffee; Infer: there is warm drink

although, *ex hypothesi,* the network's behaviour conforms to that rule.

As in the discussion of syntax, we can make essentially the same point in terms of the sentence interpretation example. A network with distributed, microfeatural, input and output encoding may achieve conformity to the rule

> Given: the sentence presented contains the predicate "is drunk;"
>
> Infer: the proposition conveyed concerns the property of being drunk.

But, if the input representation of the occurrence of the predicate "is drunk" varies from case to case, then the explanation of the network's conformity to this rule in one case will not be just the same as the explanation in another case.

None of this is to say, of course, that in connectionist networks completely distinct and autonomous processes are involved in the various transitions that accord with a pattern. Connectionist networks offer an option in between strict commonality and strict autonomy or modularity. They fall between systems with knowledge of rules, on the one hand, and mere look-up tables, on the other.

Connectionism presents no problem for the conditional claim that if there is causal systematicity of process then there is syntactic structure in the input states to that process. For, with distributed representation, there is typically neither syntax nor systematicity. What is more, there may not be systematicity of input-output process, even where there is syntactic structure in the input states.

This typical departure from causal systematicity does not, by itself, constitute an objection to the connectionist programme. If a cognitive process is causally systematic, then distributed connectionism is unlikely to provide a good model for that process. But it is, in general, an empirical question whether any given cognitive process is systematic in the sense that concerns us. Consequently it is, in general, a matter for detailed empirical investigation whether modelling actual cognitive processes presents a problem for connectionism.

However, the second stage of Aunty's argument reveals that there is a tension between the connectionist programme for modelling cognition and our commonsense conception of ourselves as thinkers. On the face of it, the connectionist paradigm does not provide a good scientific psychological model for the domain of conceptualised thought and inference.

IX. AN INVITATION TO ELIMINATIVISM?

If all this is right, then what seems to be in prospect is an argument from connectionism to eliminativism; not, to be sure, to the elimination of all semantic content, but to the elimination of the bearers of semantic content that belong in the commonsense scheme: beliefs, and thoughts in general.

The present argument for a tension between the commonsense scheme and the connectionist programme finds a parallel in the paper by Ramsey, Stich and Garon (1990). They defend a conditional claim (p. 500):

> *If* connectionist hypotheses . . . turn out to be right, so too will eliminativism about propositional attitudes.

Their argument comes in two main stages. First, they claim that the commonsense scheme is committed to propositional modularity. This is the idea that (p. 504):

> propositional attitudes are *functionally discrete, semantically interpretable,* states that play a *causal role* in the production of other attitudes, and ultimately in the production of behaviour.

Then, second, they claim that distributed connectionist networks do not exhibit propositional modularity.

The argument of the present paper is likewise an argument for an incompatibility between a feature of the commonsense scheme and connectionist hypotheses. Ramsey, Stich and Garon argue: Networks do not exhibit propositional modularity; the commonsense scheme is committed to propositional modularity; therefore connectionism is opposed to the commonsense scheme. Similarly, Aunty and I argue: Networks do not exhibit syntax and causal systematicity of process; the commonsense scheme is committed to syntax and causal systematicity of process; therefore connectionism is opposed to the commonsense scheme.

The parallel extends to some points of detail. Ramsey, Stich and Garon argue that in a connectionist network there are no functionally autonomous vehicles of proposition-sized semantic contents. In the case where the putative vehicles under consideration are patterns of weights, their point is essentially similar to the claim that processing in networks is not causally systematic. This is hardly surprising. For suppose we focus upon the role of beliefs in mediating between desires and action, or in mediating inferentially between other beliefs. Then what propositional modularity requires is that there should be functionally autonomous transition mediators. And that is also what is required if the several transitions—from desire to action, or from belief to belief—are to be causally systematic.

Each argument purports to establish a necessary condition for a being to be a thinker (a believer, a deployer of concepts). In each case, this necessary condition concerns internal cognitive architecture; it is a condition that is far from being guaranteed by facts about behaviour. And, in each case, it is claimed that a being whose internal cognitive architecture is correctly described as a connectionist network will not meet the necessary condition for being a thinker which the argument purports to establish.

If either argument is sound then we are faced with a puzzle. On the one hand, we cannot prejudge *a priori* the outcome of research in psychology and cognitive science. It is presumably an epistemic possibility that we should turn out to have connectionist networks inside our heads.

On the other hand, many philosophers will argue that it is incoherent to suppose that it might turn out to be simply false that we are thinkers and believers (see e.g. Schiffer 1990). There is a powerful intuition that it is part of our very conception of a believer or thinker that we are the paradigm exemplars. Consequently, eliminativism seems an unappealing prospect.

If we reject *a priorism* about the future development of science, but still seek a line of resistance against eliminativism, then we can explore the possibility that the commitments of the commonsense scheme may be met at some higher level of description of connectionist networks.

The argument of Section 7 established only that the requirements of syntax and systematicity are typically not met at the level of description of networks in terms of units and connections, activation and weights. That does not rule out the possibility that some analysis of the operation of a network (such as cluster analysis) may vindicate a higher level of description from whose point of view the approximate and blurred commonalities are just variable realisations of real commonalities. (See Clark 1989, 1990; Davies 1990a, 1990b.)

To the extent that the connectionist programme permits multiple levels of description of networks to contribute to psychological theory, the tension between that programme and the commonsense scheme may be reduced. But it cannot be altogether removed. For there is no guarantee that, if we turn out to have connectionist networks inside our heads, then they will be networks that meet the requirements of syntax and system (or propositional modularity) at some vindicated level of description.

We must live with the residual tension between connectionism and the commonsense scheme. And, in general, we must live with the prospect that empirical discoveries about cognitive architecture may come into conflict with our commonsense conception of ourselves. For the only way to rule out that prospect altogether is to empty the commonsense scheme of all commitments to internal architecture; that is, of all commitments that go beyond behaviour. But, if the price of relief from tension is behaviourism, then relief is something that we shall surely do without.

That residual tension between connectionism and the commonsense scheme, when combined with the epistemic possibility that connectionism might turn out to be right, leaves us with a philosophical puzzle. For it is far from easy to see how to deal with the intuition of non-negotiability, concerning the proposition that *we* are thinkers. But the resolution of that puzzle is not a project for this paper.[1]

NOTE

1. I am endebted to Katherine Morris for comments on an early version read to the Oxford Philosophical Society in November 1988; and to Ned Block, Andy Clark, and Christopher Peacocke for helpful discussions. Talks based on this material were given at the Australian National University, the University of Sydney, the University of Queensland, and LaTrobe University, during August and September 1989, and at MIT and Rutgers

University, during February and March 1990. I am grateful to ANU and the British Academy for financial support.

REFERENCES

Barwise, J. 1987. Unburdening the Language of Thought, *Mind and Language* vol. 2, pp 82–96.

Churchland, P.M. 1986. Reductive Strategies in Cognitive Neurobiology, *Mind* vol. 95; reprinted in *A Neurocomputational Perspective*, Cambridge, MA.: MIT Press, 1989, pp. 77–110.

Clark, A. 1989. Beyond Eliminativism, *Mind and Language* vol. 4, pp.

Clark, A. 1990. Connectionist Minds, *Proceedings of the Aristotelian Society* vol. 90, pp. 83–102.

Crane, T. 1990. The Language of Thought: No Syntax Without Semantics, *Mind and Language* vol. 5, pp. 000–00.

Davies, M. 1986. Tacit Knowledge and the Structure of Thought and Language. In C. Travis (ed.), *Meaning and Interpretation*, Oxford: Blackwell, pp. 127–58.

Davies, M. 1987. Tacit Knowledge and Semantic Theory: Can a Five Per Cent Difference Matter? *Mind* vol. 96, pp. 441–62.

Davies, M. 1989. Tacit Knowledge and Subdoxastic States. In A. George (ed.), *Reflections on Chomsky*, Oxford: Blackwell, pp. 131–52.

Davies, M. 1990a. Rules and Competence in Connectionist Networks. In J. Tiles (ed.), *Evolving Knowledge in Natural Science and Artificial Intelligence*, London: Pitman

Davies, M. 1990b. Knowledge of Rule in Connectionist Networks, *Intellectica* no. 9–10: D. Memmi and Y.M. Visetti (eds.), *Modèles Connexionnistes*, pp. 81–126.

Davies, M. 1991. Facing up to Eliminativism. In A. Clark and P. Millican (eds.), *Proceedings of the Turing Conference 1990*, Oxford: Oxford University Press

Dennett, D. 1990. Granny's Campaign for Safe Science. in B. Loewer and G. Ray (eds.), *Meaning in Mind: Fodor and His Critics*, Oxford: Blackwell

Devitt, M. 1989. A Narrow Representational Theory of the Mind. In S. Silvers (ed.), *Representation: Readings in the Philosophy of Mental Representation*, Dordrecht: Kluwer Academic Publishers, pp. 369–402; reprinted in W.G. Lycan (ed.), *Mind and Cognition: A Reader*, Oxford: Blackwell, 1990, pp. 371–98.

Evans, G. 1981. Semantic Theory and Tacit Knowledge. In S. Holtzman and C. Leich (eds.), *Wittgenstein: To Follow a Rule*, London: Routledge and Kegan Paul; reprinted in Evans, G. *Collected Papers*, Oxford: Oxford University Press, 1985, pp. 321–442.

Evans, G. 1982. *The Varieties of Reference*, Oxford: Oxford University

Fodor, J. 1975. *The Language of Thought*, New York: Crowell.

Fodor, J. 1985. Fodor's Guide to Mental Representation, *Mind* vol. 94, pp. 77–100; reprinted in *A Theory of Content and Other Essays*, Cambridge, MA.: MIT Press, 1990, pp. 33–39.

Fodor, J. 1987a. *Psychosemantics*, Cambridge, MA.: MIT Press.

Fodor, J. 1987b. A Situated Grandmother? *Mind and Language* vol. 2, pp. 64–81.

Fodor, J. and Pylyshyn, Z. 1988. Connectionism and Cognitive Architecture: A Critical Analysis, *Cognition* vol. 28, pp. 3–71.

Peacocke, C. 1986. *Thoughts: An Essay on Content*, Oxford: Blackwell.

Peacocke, C. 1989a. What Are Concepts? In P.A. French, T.E. Uehling and H.K. Wettstein (eds.), *Midwest Studies in Philosophy Volume 14: Contemporary Perspectives in the Philosophy of Language II*, Notre Dame, IN.: University of Notre Dame Press, pp. 1–28.

Peacocke, C. 1989b. Possession Conditions: A Focal Point for Theories of Concepts, *Mind and Language* vol. 4, pp. 51–6.

Peacocke, C. to appear. Content and Norms in a Natural World. In E. Villaneuva and L. Valdivia (eds.), *Information-Theoretic Semantics and Epistemology*, Oxford: Blackwell.

Perry, J. 1986. Thought Without Representation, *The Aristotelian Society Supplementary Volume 60*, pp. 137–51.

Ramsey, W., Stich, S. and Garon, J. 1990. Connectionism, Eliminativism and the Future of Folk Psychology. In J.E. Tomberlin (ed.), *Philosophical Perspectives 4: Action Theory and Philosophy of Mind*, Atascadero, CA.: Ridgeview Publishing Company, pp. 499–533.

Schiffer, S. 1988. *Meaning* (Second Edition), Oxford: Oxford University Press.

Schiffer, S. 1990. Physicalism. In J.E. Tomberlin (ed.), *Philosophical Perspectives 4: Action Theory and Philosophy of Mind*, Atascadero, CA.: Ridgeview Publishing Company, pp. 499–533.

Smolensky, P. 1988. On the Proper Treatment of Connectionism, *Behavioral And Brain Sciences* vol. 11, pp. 1–74.
Stich, S. 1978. Beliefs and Subdoxastic States, *Philosophy of Science* vol. 45, pp. 499–518.
Stich, S. 1983. *From Folk Psychology to Cognitive Science*, Cambridge, MA.: MIT Press.
Wittgenstein, L. 1969. *The Blue and Brown Books*, Oxford: Blackwell.

CHAPTER 10

COGNITIVE SCIENCE AND SEMANTIC REPRESENTATIONS

Jean-Francois Le Ny[1]
Universite de Paris
France

Various, sometimes largely diverging, opinions exist about cognitive science: a major contrast is between supporters of "cognitive sciences" as a plural, i.e., denoting a set of related but distinct areas, which will still remain different in the future and those of "cognitive science" as a singular, denoting a single unified domain. The first view presently seems to be easier to advocate: no complete unified body of knowledge has yet emerged across artificial intelligence, cognitive psychology, linguistics, logics, cognitive neurosciences, philosophy of mind, etc. that could both meet the commonly agreed criteria of scientific thought and be considered as concerning a unique well-defined object.

We will here take the risk, as our title indicates, to accept the second view, that of one integrated cognitive science. From this point of view, which is prospective rather than descriptive, the highly abstract concept of "cognitive science" is treated as a general working hypothesis. It is based on an undisputed observation in history of present science: development of current cognitive research shows a high degree of convergence from particular cognitive sciences. This seems to be true both on the theoretical and the empirical levels: not only are concepts often borrowed from a particular domain and introduced into another domain, showing their theoretical productivity, but also, and perhaps in a more convincing way, empirical conclusions obtained in a given science can sometimes be rephrased in the language and on the level of description of another science, thus showing the compatibility between them.

This convergence surely is far from perfect. But discrepancies in cognitive science can also be related with a lack of internal unity in the various cognitive sciences: as a matter of fact, internal differences in a particular domain may sometimes be larger than those between domains.

Thus, researchers can in practice focus on either similarities and common ideas between domains or their differences. They can choose to bet on unification as long as no decisive contrary evidence is afforded, or keep to plurality until a threshold they judge significant has been attained. We will in this paper try to show how the unitary hypothesis can be applied to the concept of representation.

Unification does not mean reduction, in the sense of elimination. Among people favoring unity in cognitive science, some go so far as to consider the most basic sciences, i.e., neurosciences or neuro-mimetic models, as able to eventually reduce all cognitive phenomena to neural-like explanations. I will in this paper prefer to put cognitive science on the ground of symbolic representation and functional processes.

I. COGNITIVE AND OTHER SCIENCES AS USING REPRESENTATIONS

Basic common aspects of knowledge aimed in cognitive science stand out if we stop evaluating internal convergences or divergences in it and rather compare this domain as a whole with other large scientific domains.

Cognitive science turns out to have a very peculiar look to its object compared to other sciences, in particular empirical ones such as physics, chemistry and biology. Whereas those other sciences have as their goal to study parts, aspects or events of the real, physical or living world, and to yield an extensive representation of it, which is potentially correct and "true," each particular cognitive science has as its goal to attend not directly to world objects or phenomena but to representations of them and modes of processing information coming from them. All of them have as their object representations of representations.

This difference, albeit very important, can be relativized indeed, if taken in the conceptual framework open by information theory, realized in computers, and expressed in a theory of extensional identity between mind and matter: i.e., if all representations of the world are considered as being themselves nothing else than parts of the real world. From this point of view, representations ultimately are states or events in human brains or computers, i.e., in the natural world. They are based on structures and correspondences that, in spite of their prodigious complexity, are nevertheless constructed from single physical components relatively common in the universe (molecules and their own constituents), communication between those knowledge structures, which is a basic property of their functioning, being also a part of the physical world.

Although this idea seems to be presently acceptable in most scientific non cognitive and cognitive areas, description of representations are actually given in several ways, with different uses and senses of the word "representation" itself.

Study of representations is under the label "representation of knowledge" a large sub-area for Artificial Intelligence. Its main goal is to yield a set of paper or machine memory structures that are easily manipulable in a certain type of processing, and able to conveniently correspond to objects and events in the physical world. The word "representation" in this phrase very often refers to nothing but this: "representation of knowledge" then is a coded and systematized representation of a domain of the real world. Only when the goal explicitly is to yield a representation of intensional representations (for example Maide and Shapiro, 1982) is the phrase "representation of knowledge" strictly used.

In this case, the goal is not different from that using cognitive psychology representations: for example intelligent tutorials may involve the necessity to construct a machine representation of the individual mental representations (in fact concepts) possessed or acquired by scholars or students, in order to make a conceptual diagnosis of these, and, if necessary, "repair" them (i.e., Ny, 1989 b and c).

Cognitive psychology gives its own description of representations, which clearly concerns mental representations. The main difficulty is then to correctly tell the individual, idiosyncratic, part of such representations —which non cognitivists both tend to falsely identify with their phenomenal aspects, and to overestimate—from the part in them that is stable, and in a certain sense, "objective," for example consonant to external reality, consistent with logical criteria, shared with other people, or, in a more restricted way, with expert people, but also has other psychological properties as familiarity, retrievability, Rosch-typicality, etc.

Such a distinction is very subtle, so that we can entirely understand some authors in the past, who took an anti-psychologistic stance when they tried to characterize conceptual representations. Let us just mention Frege as an example of this, and observe that such an anti-psychologism was a constant historical characteristic of the whole logics for centuries. Sometimes some computer scientists—who appear to not clearly see the differences between clinical and cognitive psychology—today take the same position ...and make an extensive use of folk psychology.

As a matter of fact, a general question to cognitive science is how can we elaborate for this particular type of natural entities, representations, and for their modes of functioning, an adequate, i.e.,both correct or true, and practically usable, hyper-general description, or hyper-representation, given from a natural science point of view, which correctly subsumes all other kinds of description.

II. NATURAL AND RATIONAL REPRESENTATIONS

A first step in this direction seems to make a clear distinction between two distinct, albeit not disjunctive, types of representations, here labeled rational and natural.

1. Rational representations, which some authors label "logical," are those submitted to the conventional requirements of scientific and logical thought: firstly consistency, and various inter-conceptual properties as set appartenance, inclusion, partial heritage, etc. and secondly adequacy, i.e., appropriate correspondence with the real world. Given these properties, which may be more or less strictly used, and in interaction with them, rational representations are as a rule shared by a whole human community, often of experts in a given domain: they are expressed through language in books, articles, lectures, etc. Sharing, or common acceptation, is implicitly taken as a major guarantee against subjectivity.

The word "knowledge" in "representation of knowledge" in fact refers to these rational representations. "Representation" in "representation of knowledge" in addition implies, as said above, that a coded and systematized form has been given to them, that is that the rational requirements of these representations are made explicit, for example in a standard logical way, and strictly applied in order to make them representable to the second degree in a knowledge base, and usable in a computer.

2. Natural representations are by contrast those that reside in human minds, or brains, and are consequently submitted to psychological and neurobiological laws or rules. Thus, "natural" is here an equivalent of "mental," as long as this word encompasses not only phenomenal events or entities in a mind but as well non conscious ones. Ultimately, "natural" is thus an equivalent of "cerebral," even though neurosciences presently

have no means to really assess the neural nature and connections of particular representations.

How are these two types of representations interrelated? We will in this paper adopt a radical postulate: that no representation in the world exists but centered in these type 2 representations: i.e., as either a representation-in-a-mind, or a generator of such a representation, or a copy of such a representation.

Pictorial or symbolic representations, for example a photograph, a drawn triangle, a sentence, an equation, all have as their main property and function to be external generators, in interaction with mnesic long term capabilities, of mental events having the form of representations in humans. Machine representations are first degree copies of representations-in-the-mind of the system's conceptor, and sometimes higher degree copies of other representations, for example a knowledge rule in an expert system is a $copy_3$, expressed in a given format, of a $copy_2$ in a cognitive engineer's mind of a $representation_1$ in an expert's mind, that is: of a $copy_1$ of a given relationship in a sub-domain of the world.

It is highly productive, from a scientific cognitive point of view, to consider rational representations, and their machine copies, in this way, i.e., as a derived, rationalized subset from natural representations.

Such a subset is obtained by adding certain properties to natural representations, or, more precisely, by laying down additional restrictions on their natural attributes. For example, in an individual two distinct natural representations in a same semantic field may happen to have anyone of two possible values, positive or negative, on their attribute "*consistency*"; rational representations always must have the value "positive consistency."

Similarly, natural representations that have the role of sub-concepts of a concept may have various values on their attribute "*Rosch-typicality*" (or "*representativity*") in respect to their super-concept for example representations of "sparrow" are more typical of "bird" in most subjects than those of "ostrich." But rational sub-concepts of a concept have no such property as a distinct typicality value, that is they necessarily have one and only one value[2] for appartenance or inclusion, restricted to 1.

In this view, concepts are particular contents (not forms) of representations, and have no independent existence outside minds only the objects, qualities, events, relationships that are their extensions can exist in such an independent way. In particular, abstract representations as mathematical ones are by no means copies of "another" category of entities, existing in a different world of ideas. They are a subset of natural representations, formed in minds as a function of both: 1. definite conditions found in the real world, 2. explicit requirements for rationality, accepted by humans. We will here not elaborate this position as concerns the foundations of mathematics.

It may of course happen as another consequence of the previous view that some supposedly rational representations keep, to some degree, natural characteristics that make them incompletely rational.

III. SOURCES OF VARIABILITY IN REPRESENTATIONS

We can now try to present a model of this that emphasizes sources or modes of variation in representations. Let us take as an example the representation expressed by the phrase: (1) "*the discovery of America by Christopher Columbus.*" Italics here indicate that we are dealing with the representation, not the fact.

From a linguistic point of view, it is a composed phrase, a nominalization of the sentence: (2) "Christopher Columbus discovered America." This sentence is true, since an event corresponding to it existed in the history of the real world, so representation (1) is adequate. We can say, in a way that is not actually conventional (see Le Ny, 1989a for a discussion) but apparently acceptable by everybody, that the fact is the *extension of this representation*.

Let us now write representation (1) as: (3) R(e,A,P), in which e is a constant, the extension of the representation as defined above, i.e., the unique fact described in (2). A denotes the set of possessors of this representation, i.e., any cognitive agent in which the representation of e can reside, which we will be concerned about in a moment. P denotes the properties of the representation, that will be dealt with in a following section.

We know that representation (1), i.e., with extension e, is possessed by a number of humans. In addition, machine representations can be created in various forms, the semantics of all these being e, "the discovery of America by Christopher Columbus." Thus A, as a variable denoting the entire set of possessors, can be split up in two subsets, that of humans and that of knowledge bases containing this representation.

We are thus led to consider R not as a constant denoting a unique representation, but as a variable denoting a class of them, in this case identified by their common extension e. This class contains a number of instance representations R, that vary in various ways according to their possessors, conditions of formation, times of activation, etc. The possessors seem to be the main source of variation in this class, that is, in addition to the gross difference between men and machine, on the one hand variability between humans, or several sub-sets of them, defined by their ages, geographical origins, levels of instruction, etc., and on the other hand variability between knowledge bases and machine supporting them.

But the interesting question is: how can we appropriately characterize these differences, that is which properties belonging to R, abstractly summarized under P, must we take into consideration in order to appropriately describe all these instance representations. In other words, which are the common properties and relevant modes of variations of the representations expressed as a function of the nature and identity of their possessors?

Several types of responses to this question are possible, which are deeply distinct. As a matter of fact, each particular cognitive science gives its own description of representations, and their properties, at its own level of description.

It is interesting to first examine the use of physico-chemical descriptions: they primarily concern the stuff and basic neurobiological modes of functioning of representations and the contrast they lead to. As a matter of fact, this source of variation can only be now very general: even though we can adopt, at least as a working hypothesis, the view that any representation has its material realization in the functioning of a cell assembly, no method permits to observationally distinguish between a given representation and another representation, both being individualized by a cognitive property, for example their extension.

At the very general level, if we did consider this physico-chemical characterization as important we could add a special index to our notation, for example C in: R (e, A, C, P) which would denote the material of which representations are made.

Such a source of variation would presently oppose c1 (=made of silicium, including the physical phenomena and properties associated with semi-conductors and integrated circuits), and c2 (= made of carbon macromolecules, including the physico-chemical phenomena present in individual neurons or synapses, the role of specific neuro-transmitters, formation of neural assemblies, etc.). No doubt we will have in the future to add properties as c3 (= optoelectronic), c4 (= neuromimetic parallel hardware and architecture), and others.

The present characterization generates divergences in cognitive science. Differences in structure and functioning between a human brain and a present computer are known to be very large and to show a "thrilling contrast" (Paillard, 1987): although they did not prevent symbolic Artificial Intelligence to grow, some researchers think that future advances in cognitive science can only be obtained if the structure of computers is made closer to that of the brain.

One principle often focused by advocates of this view is the basic biological principle that "structure determines function." From this, the idea derives that neurosciences can yield the right level of description of cognitive representation and functioning (see Changeux, 1983), and guide creation of new kinds of processing machines.

Undoubtedly development of new basic components and architecture in computers, in particular parallelism, can have a very large impact on cognitive science. In particular, a very important possibility they can bring is improvement of the capacities of learning in artificial systems, which presumably can best permit elaboration of representations as rich as those humans.

However, we cannot believe that presently unsolved problems in cognitive science will find a solution in hardware improvement, and in modeling the structure of the brain. Contrary to the general principle quoted above, an extraordinary change substantiated by cognitive science is precisely the use of the inverted principle that, at least in symbolic cognition, "function determines structure."

In other words, since learning and organized memory are the basis for the structure of representations, psychological research shows that the early knowledge structure of the mind is so largely undetermined that its contents can only be formed in personal history after the structures of the world: the basic principle of empiricism (which we can take as only partially valid, leaving a part of innate general structures and processes) must be substituted to the basic biological principle as concerns cognition. Assumption that both structural indetermination and capacity of progressive structuring is also a main property of the higher part of the brain, which leads to a functionalistic view of cognitive science, seems to be more compatible with the facts.

Then, a functionalistic view of cognitive science, in particular as concerns representations, is far more acceptable that the dream of a neurobiological reductionism, which would replace and eliminate other sciences in this area. Below we will examine problems of the compatibility of symbolic representations with their neurobiological substrate.

IV. USE OF PRESCRIPTIVE RULES

If we disregard material properties of representations, we are led to their other properties, which can be natural or rational.

It has been argued above that ordinary logics, its use in artificial systems of information processing and knowledge representation, are precisely based on the requirement to diminish variability of natural representations. Many logical enterprises in history were motivated by such a will to supersede variability of individual beliefs by adopting regular systems of thought.

The means was adoption of prescriptive rules and their application to natural representations. These logical rules are often presented as rules of reasoning, but they often apply to representations themselves. They are of the general type: "check your spontaneous activities of information processing and automatically generated concepts, and, if necessary, restructure them, giving up the inappropriate ones, so as to comply to general principles, historically selected by successive communities of experts to govern rational thought."

The last decades have shown that these rules can be formalized not only for work "on paper," but also for material implementation and operationalization in computers: these make the informational structures of data, assertions and inferences materially manipulable through use of the physical phenomena taking place in the machine.

It seemed also easier in this case to first consider these prescriptive rules as "logical," and applied to machine deduction and "reasoning." However, development of "knowledge representation" as a related field shows that prescriptive rules need to be applied to the structuring of information as well as to its processing. Introduction of non-monotonic logics (McDermott, 1982, McDermott, Drew and Dale, 1980), whose role is to permit correction of derived knowledge when the early assertions or beliefs turn out to be inappropriate, is one more illustration of this interaction between structure of information and reasoning "salva veritate."

V. DESCRIPTION OF NATURAL REPRESENTATIONS

We can now examine some properties of natural representations. For cognitive psychology, the best view seems to consider that the word "representation" in fact covers two distinct kinds of entities: token representations, which are cognitive events, and long term memory representations, which are cognitive structures. The type/token distinction is borrowed from linguistics, in which it primarily concerns words: the corresponding concepts can be cautiously applied to the present domain.

In linguistics, "token word" is perfectly clear and denotes a single occurrence of a given lexical unit in a context. "Type word," which seems to be immediately understandable, in fact is the product of a double generalization across the various occurrences: the first generalization is across all occurrences in a given text, or speaker, the second across all texts, or speakers.

If we transpose this distinction to the domain of mental representations in cognitive psychology, we must keep in mind that we use several separate generalizations.

Token representations can also be clearly defined in this framework. They are single mental events taking place in the mind of an individual. A particular token representation, supposed to be individuated by its content c_i, is any mental event in any individual during which this individual either thinks of c_i, or imagine c_i, or speaks about c_i (i.e., produces speech about this representation, previously present in his or her mind), comprehends c_i (i.e., constructs this representation in mind as a result of audition of words and appropriate mental activity of comprehension), etc. Such a particular token representation can be written out as $r_i(c_i, b_i, p_i)$, in which b_i is an individual, and p_i the particular properties associated with this representation.

What is c_i in such a description? If we restrict ourselves to the easiest problem, that of representations that are verbalizable, we can use the same example as above, that of the content associated with the phrase: "*the discovery of America by Christopher Columbus.*"

However, it is not very satisfying for natural representations to have their individuation only based on their extension, as they were up to now. We certainly would prefer to rather individuate them by their "psychological content" or "intention." It seems that this can be best done, in a natural science framework, by resorting to the psychological process of recognition. It is a common behavioral evidence in subjects that they can have a "same" representation again and again: the criterion of "sameness" then clearly is that subjects recognize these events as tokens of a same personal type. Recognition thus is the basis for generalization in a given subject across individual token representations with the particular content c_i. This content, taken as a variable C, of course is one of the properties P mentioned above.

If we now want to generalize across individuals, and say that several persons have a type representation that is "the same" with respect to its content, we cannot escape all the current philosophical problems concerning reference (i.e. relationship between e and c, in our notation). However, as we are not here primarily interested in this kind of problems, we will rely on the overall observation that in everyday life a close correlation exists between content and extension.

Our major concern is to characterize a type mental representation as being a class, R (c_i, A, P)—supposed to correspond to R (e, A, P)—of all mental events that are in all individuals about the content c_i, and correspond tot he event e this is exemplified by "THE representation of *the discovery of America by Christopher Columbus,*" considered in all its realizations in all subjects. In this analysis, when we say: "THE concept of *flower,*" we are exactly mentioning such a class. Thus we are now prepared to focus scientific research on the natural properties associated with particular (sub-classes of) representations in subjects.

Which kinds of such properties should we take into consideration? We already mentioned psychological properties as familiarity, accessibility from memory, Rosch-typicality, etc, that are as a rule applied to type representations, and object of many researches.

We must separately consider other properties as belonging to token representations: for example those created by the context in which a representation content is used.

Let us illustrate this via an example: suppose that the possessor s of the representation "the discovery of America by Christopher Columbus" is a scholar attending a lecture about history of American Indians and their peculiarities in the end of the XVth century. Now suppose that s was just hearing a sentence as: "presumably some visitors came from Europe to the American continent before the discovery of America by Christopher Columbus in 1492." We can admit that the representational unit "*America*" will be highly activated in s mind (see below), whereas the unit "*Christopher Columbus*" will be rather less. By contrast, if s was attending a lecture about the life and history of Christopher Columbus himself, and hearing a sentence such as: "the discovery of America by Christopher Columbus was a great success for him, although...,"we can admit that the representational unit *Christopher Columbus* will be more activated, whereas that of "*America*" will be less. Experimental evidence, based on measure of response times, support this view (for example Denis and Le Ny, 1982; Le Ny, 1990).

A related important property of token representations must now be mentioned. It can be expressed by the following question: is a particular token representation a replication obtained from a generic matrix representation, or the unique result of a non reproducible construction generated from particular parts? The first alternative involves a "realistic" use of the idea of "type representation" that will be elaborated below.

We deliberately chose to illustrate this paper by a large representation, expressed in a complete phrase, "*discovery of America by Christopher Columbus*," rather than a simple conceptual word as "*flower*," which could have introduced some unseen bias in our analysis. In addition we chose to use as the extension of this representation an event (discovery) rather than a class of objects (as "*flower*" or "*bird*" might have made). The reason of this choice was that we tried to avoid typical, hackneyed examples that are two often used in cognitive literature.

Our example of representation is clearly composed: linguistically, it is a description, as would be the meaning of "the discovery of the Desolation Islands by Kerguelen." It is initially constructed from parts, i.e.,the nominalized predicate "discovery-discovery"and its two arguments "Desolation Islands" and "Kerguelen." We will assume in the following that the representation involving America is highly familiar in most subjects, whereas the one about the Kerguelen Islands is very unfamiliar.

Our claim is then that these two representations have distinct natural properties when they are realized as tokens. The first has a counterpart in long term memory, so that any corresponding token of it is an *activation* of such a *long term memory representation* whereas any token of the second is the result of a semantic *construction involving predication*. An alternative view, which we will not here discuss, is that all token representation are constructed in the same way.

VI. TOKEN REPRESENTATIONS, LONG TERM MEMORY REPRESENTATIONS, AND THE NOTION OF ACTIVATION

The view adopted above implies a distinction between "type representations," which is an abstract notion similar to "type word," and "natural basis of token representations," which involves psychological and, possibly, neurobiological assumptions about the *"reality of"* these representations.

On this basis, a modelization of the natural functioning of representations is made possible. The basic assumptions (see for example Anderson, 1983; Le Ny, 1989a, 1990) are:

1. there exist in long term memory lasting representations; the most typical of them are generic lexico-conceptual representations as *"flower"* or *"bird,"* but some others are not less interesting: these can have various forms in language, for example definite descriptions as *"the discovery of America by Christopher Columbus."* But some other may not necessarily involve constructive procedures: an example of this is, in France, the representation generated by the couple definite article noun: "le Débarquement" ("THE Landing"), meaning "the Landing of the Allied in Normandy in June of 1944": such representations work in the same way as proper names, but their main property seems to be their high familiarity, which makes them easily accessible from memory; other, encyclopedical, knowledge can be lastingly stored with them;

2. a basic property, activability, and a basic process, activation, are associated with these long term memory representations; they make these representations permanently having a variable *level of activation* which is created by information processing. Above a given level of activation, a token representation is generated in working memory, which can thus have various levels of super-threshold activation; several behavioral measures (as processing times) can be used as indices of these changes and states of activation (Le Ny, 1990);

3. multiple relationships between representations, which can be modeled in the form of a semantic network and spread of activation between representations as a function of these relationships are additional assumptions, which will not be elaborated here.

These three main assumptions define the family of models based on symbolic representation and activation, which are among the best used for explanation of various cognitive processes, in particular language comprehension. As a matter of fact, these symbolic models are fully compatible with sub-symbolic models, based on parallel distributed processing and use of so-called "neural networks" (MacClellend and Rumelhart, 1986; Rumelhart and MacClelland, 1986). The open question between these is whether the concept of symbolic representation is absolutely needed in cognitive science or not: our position in this text is clearly affirmative.

On the contrary, the activation process assumption here advocated is different from the computer-like information processing and working memory models, according to which processing is based on transfer from one memory store to another or several memory stores. We will not elaborate this second family of models either.

One advantage of the present view is that it makes modelization of construction of new complex token representations easy. This can best be illustrated in language comprehension, in reading a sentence, a paragraph or a text. Construction of meaning is then assumed to be based on multiple activations of several representations corresponding to several elementary words: each stimulus word brings about activation of its own meaning, i.e., the semantic representation associated with its morphological representation in long term memory. From this, connection of several meanings in a unique meaning proceeds by predication and binding.

Predication is assumed to play a major role in this integration: this is a consequence of the so-called "propositional assumption," which we prefer to label "predicative assumption." According to this, elementary meaning units (corresponding to words) are stored in memory in a form that make them counterparts of logical predicates: basically, these representations own a particular type of associativity with other units.

Let us take again as an example the meaning stored with the word "discovery." The predicative assumption states that this meaning involves: 1. knowledge that there is necessarily a "discovery of (x) by (y)," 2. complementary knowledge of what x and y possibly denote, for example for x a set of geographical places, and y a set of humans; 3. more or less clear knowledge of the probabilistic distribution of various sub-sets of x and y, with their bonds.

Comprehension of larger units, as complete sentences, paragraphs or texts, may be modeled on this basis as establishment of other bonds between propositional or super-propositional units or constituents, corresponding to various kinds of schemata in memory (for example Yekovich and Walker, 1986, 1987). Similarly, units of meaning can be viewed as representational parts that are not necessarily attached to a whole morphological units in a one-to-one way, but as well as comprising "smaller-than-the-world" components: this is the basis for a componential view that would escape the criticisms developed against previous feature theories.

Descriptions of such semantic representations subsequent to comprehension can be experimentally tested in an indirect way by comparison with several kinds of empirical data, in particular those collected on response times in experimental standard situations, as self-paced reading, probing or priming. Present evidence is in a not too bad agreement with these models more results must be collected in order to be sure that they are valid, and which particular model is the best. But internal compatibility, i.e., compatibility between models and facts is undoubtedly growing.

VII. CROSS-COMPATIBILITY WITH NEUROBIOLOGY AND ARTIFICIAL INTELLIGENCE

As the family of symbolic models based on the three notions of token representation, long term memory representation and activation has an acceptable validity inside cognitive psychology, the further problem is whether it can be also considered as compatible with concepts and knowledge based on facts collected by other methods, and having a high plausibility in other cognitive sciences. This question derives from the rule to circulate concepts and explanation across areas and methods. It is of particular concern for Neurobiology and Artificial Intelligence.

The idea of activation is itself borrowed from neurobiology, although its second argument—"activation of (what?)"—has a deeply different semantics according as it is applied to the domains of neural or symbolic entities. However, the related idea of token representation can also easily find a neurobiological counterpart when viewed as an event. The psychological model of activation is congruent with everything we know about a category of neural events: change of a neural configuration from an inactive (or few active) to a more active state.

Generally speaking, activation of specific neural assemblies is today the best neurobiological candidate for a support of representational activity, even though neural events that can now be evidenced only concern too small or too large parts of the nervous system, and in no way something that could be identified as a specific and relevant neural assembly.

However, there exists experimental data (see for example Block and Laroche, 1986) showing concomitance between well-specified neural events and definite behaviors, as a rule behaviors of anticipation, permitting to infer presence of representations in animals. It is, and certainly will remain, more difficult to say of what these events may be representations. It is interesting to observe that the philosophical problem of "inscrutability of reference" (Quine, 1961) here takes a new form.

There is more doubt about a possible counterpart of lasting representations. In some versions of parallel and distributed processing, in which the focus is put on "blind"processing, there is no place for such an idea, or even negation of it. But otherwise, it may be easily assumed that the reality of long term representations resides in the sub-network structure of the connections in a neural pattern. The working hypothesis we use here is thus compatible with the idea of such "neuro-symbolic" patterns, which would be specific counterparts of cognitive lasting structures, meaning units or concepts.

In short, a fairly satisfying compatibility can be found between this family of psychological models and facts, and those of neurobiology. From this, two-way constraints can continue to be imposed on these models by future facts.

A different kind of compatibility must be looked for on the Artificial Intelligence side. We said above that we do not believe it could be found directly between A. I. and neurobiology, at least as concerns high level phenomena: this compatibility must go through cognitive psychology and functional identities at a common level, the symbolic or "knowledge level" advocated by Newell (1982).

This can again be best illustrated by problems concerning natural language processing, in particular those that are common to automatic and natural comprehension.

Most current systems of language interpretation have as their goal to construct, from information given as an input text, a final semantic (or syntactico-semantic) output machine representation. This must both 1. permit subsequent processing, for example production of responses in question answering, summaries or indexing, performing a dialogue, etc.; 2. yield a "representation of the text's meaning" that is fairly acceptable by an expert.

As a final semantic, mental representation is also the result of human language comprehension, "fairly acceptable" in the previous sentence should be taken as referring to a description of this semantic human representation rather than a particular linguistic theory.

It could be shown in more detail than here possible that a set of similar functions must be accomplished, either by the machine automat or by the human mind, to attain this goal. Let us briefly mention. 1. morphological identification of the successive various units (words) contained in the text; 2. access to (or activation of) the meaning associated with the types (roots and flexions) of these forms, taking into consideration, if necessary, the context of these units; 3. predication into the propositional format, as already mentioned above; 4. syntactic analysis; 5. identification and filling up of semantic and grammatical schemata stored in long term memory (as frames, scripts, rules of word order, etc); 6. storage in separate registers of small partial representations (semantic counterparts of constituents); 7. further binding of these meaning parts into larger parts, with progressive transformational of partial representations into a whole representation; 8. decay and particle loss of information, as a function of its importance; 9. use of reprocessing to modify representations and estimation of importance; etc.

The two lastly mentioned functions, which emphasize the role of working memory and selection of information in comprehension, are not often mentioned in literature, but we consider them as very important (Le Ny, 1990a,b).

The ways these functions are realized in mind and machine are very different, but they can be considered as highly compatible. For example, instantiation is a high-level basic mechanism for language processing, of which activation can be considered to be a natural form, whereas copy and adjunction of information is a form in computer.

In short, language comprehension can be viewed as a set of transformations applied to representations, having their end in a final comprehensive representation. A specific fact concerning human comprehension is that this final representation has the very particular property " conscious," whereas either intermediate human or machine representations have not.

VIII. CONCLUSION

We tried in this paper to show how an orientation can be taken in cognitive science, in which exchange between concepts, models and facts would permit to mutually enrich current knowledge. Certainly, very much remains to be done before we can construct a "hyper-general," i.e., a high-level, superordinate, description of representations and information processing. But this effort of circulation of ideas is both a necessary and possible goal in the quest of an integrated cognitive science. Anyway, a more complete elucidation of the general concepts in this large area, in particular of "representation," is a valuable means for elaboration of hypotheses and directions of research useful for empirical work.

NOTES

1. Institut des Sciences Cognitives et de la Communication Université de Paris-Sud. Centre Scientifique d'Orsay, Bat. 335. 91405 Orsay Cédex - France.

2. We are here implicitly taking a position against fuzzy set logics. See Le Ny, 1989a for an analysis.

REFERENCES

Anderson, J. (1983). *The architecture of cognition*, Cambridge, Mass.: Harvard University Press.

Block, V., and S. Laroche (1986). Facts and hypotheses related to the search for the engram. In G. Lynch, J.L. McGaugh and N.M. Weinberger (eds). *Neurobiology of Learning and Memory*. New York: Guilford, pp. 249-260.

Changeux, J. P. (1983). *L'homme Neuronal*. Paris: Fayard.

Denis, M. and J.F. Le Ny (1986). Centering on figurative features during the comprehension of sentences describing scenes, *Psychological Research,* 48, pp. 145-152.

Le Ny, J. F. (1989a). *Science cognitive et compréhension du language*. Paris: Presses Universitaires de France.

Le Ny, J. F. (1989b). Automatic induction of concept contents from verbal characterizations. In G. Toberghien (ed.) *Advances in cognitive science*, vol. 2. Chichester, N.Y.: Harwood-Wiley, pp. 165-177.

Le Ny, J. F. (1989c). The role of schemata in concept acquisition and diagnosis. In H. Maurer (ed) *Computer assisted learning*. Berlin: Springer Verlag, pp. 308–321.

Le Ny, J. F. (1990a). Compréhension du language, représentations sémantiques en mémoire de travail et processus du reconnaissance, *Communicazioni scientifiche di psicologie generale*, in press.

Le Ny, J. F. (1990b). Coherence in semantic representations: text comprehension and acquisition of concepts. In G. Denhière & J. P. Rossi (eds) *Text and text processing*. Elsivier-North-Holland, in press.

McClelland, J. I., D. F. Rumelhart and the PDP Research Group (1986). *Parallel distributed processing explorations in the microstructure of cognition, vol. 2: psychological and biological models*. Cambridge, Mass: MIT Press.

McDermott, D. (1982). Non-monotonic logic II Non-monotonic modal theories, *Journal of the ACM*, 29, pp. 33–57.

McDermott, D., v. Drew and J. Doyle (1980). Non monotonic logic 1, *Artificial intelligence*, 13, 41–72.

Maida, A. S. and S. C. Shapiro (1982). Intensional concepts in propositional semantic networks, *Cognitive science, 6*, 291–330.

Newell, A. (1982). The knowledge level, *Artificial intelligence, 18*, 87–127.

Paillard, J. (1987). L'ordinateur et le cerbeau: un contraste saisissant, *AFCET/Interfaces*, 57, pp. 4–9.

Quine, W. V. (1961). *From a logical point of view*. Cambridge: Harvard University Press.

Rumelhart, D. E., J. L. McClelland and the PDP Research Group (1986). *Parallel distributed processing explorations in the microstructure of cognition, vol. 1: Foundations*. Cambridge, Mass.: MIT Press.

Yekovich, F. R. and C. H. Walker (1986). Retrieval of scripted concepts, *Journal of memory and language*, 25, pp. 627–644.

Yekovich, F. R. and C. H. Walker (1987). The activation and use of scripted knowledge in reading about routine activities, in B. K. Britton (ed), *Executive control processes in reading*. Hillsdale, N.J.: Erlbaum.

CHAPTER 11

ANCHORING CONCEPTUAL CONTENT: SCENARIOS AND PERCEPTION[1]

Christopher Peacocke
Oxford University
England

I will be developing a suggestion about the way in which perceptual experience represents the world. I want to explore the consequences of this suggestion, and to apply it in addressing various questions about the relations between perception and the conceptual content of thought. These concerns set the itinerary for this paper. But there are also intriguing regions adjacent to the main route. I will indicate these as we go, especially in the later sections of this paper. It seems to me that our understanding of these issues is still extraordinarily primitive. I hope the suggestion I will be developing will lead to new routes into this still largely unmapped territory.

I. SCENARIOS INTRODUCED

I will be concerned with a subset of the representational properties of experiences: that is, with a subset of the properties of an experience in virtue of which the experience is assessable as correct (veridical) or incorrect in the way it represents the world as being. I'm not going to be concerned with all such representational properties. In particular, I am not going to be immediately concerned with those properties that depend upon the perceiver's abilities to recognize individual material objects, or to recognize an object as, say, a chair or a dog. I would argue, though, that the subset of properties I will be discussing is fundamental, in that the nature of representational properties outside the subset is to be explained at least in part by their relations to those inside the subset. Perhaps it is best to begin by giving a rough, intuitive statement of the thesis for which I shall be arguing.

The thesis is that a representational content of the sort in question should be individuated by specifying which ways of filling out the space around the perceiver are consistent with the correctness of the given representational content. So the content involves a spatial *type*, the type being that under which fall precisely those ways of filling the space around the subject which are consistent with the correctness of the content. On this model, correctness of a content is then a matter of instantiation: the instantiation by the real world around the perceiver of the spatial type which gives the representational content in question.

This account is potentially of importance to anyone interested in anchoring notions of conceptual content in some level of nonconceptual content. For a spatial type is just that—the notion of a type under which certain segments of the world may fall. A spatial type is quite different from a concept. The identity of a concept, as the term is used here, is answerable to Fregean considerations of cognitive significance. A concept is also ultimately individuated by the condition required for a thinker to possess it. A spatial type is not. So a theory of nonconceptual content which employs the notion of a spatial type provides one way in which a hierarchy of families of concepts can be grounded in a noncircular way.

Now let me sharpen up the intuitive formulation. There are two steps which we have to take if we are to specify fully one of these spatial types. The first step is to fix an origin and axes. The origin and axes will not be a specific place and set of directions in the real world. This is precisely because at the moment we are fixing a type which may potentially be instantiated at many different places in the real world. Nevertheless, it is important that the origin and axes be *labelled* by certain interrelated properties. It is this labelling by interrelated properties which helps us to constrain what are instantiations of the spatial type we are determining. Thus, for instance, one kind of origin is given by the property of being the centre of gravity of a human body, with the three axes given by the directions back/front, left/right and up/down with respect to the centre of gravity (in the standard upright orientation of the body). The use of a particular set of labelled axes in giving part of the content of an experience is not a purely notational or conventional matter. The appropriate set of labelled axes captures distinctions in the phenomenology of experience itself. Looking straight ahead at

Buckingham Palace is one experience. It is another to look at it with one's face still towards the Palace, but with one's body turned towards a point on the right. In this second case, the Palace is experienced as being off to one side from the direction of straight ahead—even if the view remains exactly the same as in the first case.

In giving the content of tactile experience, we would sometimes have to use as origin something labelled with the property of being the centre of the palm of a human hand, with axes defined by relation to parts of the hand. Actually in the specification of the representational content of some human experiences one would need to consider several such systems of origins and axes, and would need to specify the spatial relations of these systems to one another. There are many other complexities too; but let us keep things simple at this stage.

Having fixed origin and axes, we need to take the second step in determining one of our spatial types, viz., that of specifying a family of ways of locating surfaces and their properties in the space around the origin. In picking out one of these ways, we need to do at least the following. We need, for each point (strictly one should say point-type) identified by its distance and direction from the origin, to specify whether there is a surface there, and if so what texture, hue, saturation, brightness and temperature it has at that point, together with its degree of resistance to touch. The orientation of the surface must be included. So must much more in the visual case: the direction, intensity and character of light sources; the rate of change of perceptible properties, including location; indeed it should include second differentials with respect to time where these prove to be perceptible.

There is no requirement at this point that the conceptual apparatus used in specifying a way of filling out the space be an apparatus of concepts used by the perceiver himself. Any apparatus we want to use, however sophisticated, may be employed in fixing the spatial type, however primitive the conceptual resources of the perceiver with whom we are concerned. This applies both to the apparatus used in characterizing distances and directions, and to that employed in characterizing the surfaces, features and the rest. I will return to the significance of this point later.

We are now in a position to say with slightly more precision what one of our spatial types is. It is a family of such ways of locating surfaces, features and the rest in relation to such a labelled origin and family of axes. I call such a spatial type a *scenario*.

With this apparatus, we can then say what is required for the correctness of a representational content of the sort with which I am concerned. Consider the volume of the real world around the perceiver at the time of the experience, with an origin and axes in the real world fixed in accordance with the labelling in the scenario. We can call this a *scene*. The content of the experience is correct if this scene falls under one of the ways of locating surfaces and the rest which is in the family of ways in the scenario.

It is important to give for experiences a notion of their representational content which is evaluable as correct or not outright, rather than merely as correct or not relative to some assignment or other. The point parallels (and indeed is connected with) a familiar point in the philosophy of language. Consider a particular utterance of the indexical sentence "He is witty." A theory should provide a statement of the conditions under which this particular utterance is true outright, rather than merely the conditions under which it is true relative to an assignment of objects to its indexical elements.[2] It is the content of the utterance which is assessable outright which concerns particular objects. Similarly, it is the content of a perceptual experience which is assessable outright that concerns particular places. For perceptual experience, I identify such an outright-assessable content with a *positioned scenario*. A positioned scenario consists of a scenario, together with (i) an assignment to the labelled axes and origins of the scenario of real places and directions in the world which fall under the labels, and (ii) an assigned time. For a particular perceptual experience, the real directions and places assigned at (i) are given by the application of the labels to the subject who enjoys the experience—if the origin is labelled as the centre of gravity of the body, the real place assigned to it is the centre of gravity of the perceiver's body; and so forth.[3] The time assigned at (ii) is the time at which the perceptual experience occurs. We can then say that the content given by the positioned scenario is correct if the scene at its assigned places falls under its scenario at the assigned time, when the scenario is positioned there in accordance with the assigned directions.

I should emphasize that the positioned scenario is literally meant to be the content itself. It is not a mental representation of the content; nor a specification of a representation; nor a way of thinking of anything. A particular component of a scenario content may be captured by many different mental representations. For example, the orientation of a surface at a particular point may be given by specifying its slant and tilt, or equally by specifying components for its representation in gradient space.[4] Mental representations which differ in the way such orientations are represented may nevertheless represent the same scenario as instantiated in their respective subjects' surroundings.

Since we have now touched on the issue of mental representation, it may be helpful for me to comment on the relation between the account being developed here and Marr's 2-1/2D sketch. There can be much in a positioned scenario which is not in the content of a 2-1/2D sketch. The 2-1/2D sketch has only retinocentric coordinates, and does not include illumination conditions. The present material does, though, give a natural framework in which to give the content of a 2-1/2D sketch, if such mental representations exist. The major question, however, is whether the present account is committed to the existence of mental representations with roughly the properties of Marr's 2-1/2D sketches. In Marr's work, such representations are computed after the primal sketch and before any 3-D models are assigned to shapes in the environment. Is the present philosophical treatment committed to the existence of such a temporally intermediate representation? It is not. What matters for the purposes to which this apparatus will be put is that there exist some mental representation at least part of whose content is given by the positioned scenario. It does not matter if the representation also has other, perhaps simultaneously computed, contents. It *will* matter that certain systematic connections hold between these scenario-involving contents and other contents that the representations may possess. I will be arguing that the identity of certain other contents depends upon the nature of their links with scenario-involving contents. But again, this does not require the existence of distinct mental representations with roughly the properties of Marr's 2-1/2D sketches.

A good theory must provide correctness conditions for perceptual experiences if it is adequately to distinguish these experiences from states which do not represent to the subject the world as being a certain way. But the importance of providing a representational content goes far beyond the need to draw that distinction in the right place. By perceiving the world, we frequently learn whether a judgement with a given conceptual content is true or not. This is possible only because a perceptual experience has a correctness condition whose holding may exclude, or require, the truth of a conceptual content. Correlatively, the account of mastery of certain concepts must exhibit a suitable sensitivity of a thinker's judgements containing them to the way his experiences represent the world to be. I will give some detailed examples later. More generally, like representational contents of other kinds, the positioned scenario content of perceptual experience makes its subject's actions intelligible. The detailed spatial actions of a subject are commonly made intelligible by the way his immediate environment is represented as being by his perceptions.

The requirement that any perceptual experience has a correctness condition imposes restrictions on what other forms of perceptual experience, besides those of human beings, are possible. Suppose it were said that there could exist a being whose perceptions have scenario-involving contents which concern an origin, but do not contain labelled axes. Now a correctness condition for a particular experience occurring at a given time is not fixed until its scenario is *positioned* in the real world. An origin alone does not suffice for this positioning: there will be many different ways of orienting it around the perceiver's location at the time of the experience. Until one of these orientations has been selected as being the appropriate one, no correctness condition has been determined. This does not rule out the possibility of spherical organisms, existing in a fluid, and whose perceptions are caused by the impact of light all over the surface of their bodies. What the requirement does mean is that, even in this case, the scenarios presented by such organisms' experiences must have axes labelled by parts of their bodies. The parts mentioned in the labelling need not be limbs. I will return to this labelling later, too.

The notion of a positioned scenario I have been employing is one that can give the content of a fully perceptual experience, and can equally give the content of an experience which is hallucinatory. But it would be quite consistent with the apparatus I've introduced to hold that the fully perceptual case has a philosophical primacy, and that non-perceptual cases have to be elucidated by the relations in which they stand to the fully perceptual case. Consider two different scenes, the objects in each of which are distinct, but which are perceived fully veridically, and in the same way. We can regard a scenario as being the type which captures the similarity of two such different perceptual cases. It is then open to us to say that a hallucinatory experience represents the environment as being a scene of such a type, though there is no such scene there. In brief, the scenario account neither exacerbates nor by itself solves epistemological problems.

II. SCENARIOS: CONSEQUENCES AND COMPARISONS

There are several desirable consequences of the thesis that the objective content of an experience is given by its positioned scenario.

(a) Writers on the objective content of experience have often remarked that an experience can have a finer-grained content than can be formulated by using concepts possessed by the experiencer. If you are looking at a range of mountains, it may be correct to say that you see some as rounded, some as jagged. But the content of your visual experience in respect of the shape of the mountains is far more specific than that description indicates. The description involving the concepts *round* and *jagged* would cover many different fine-grained contents which your experience could have, contents which are discriminably different from one another.

This fine-grained content is captured in the scenario. Only those ways of filling out the space around you which are consistent with the veridicality of your experience will be included in the scenario. The ways included in the scenario will omit many which equally involve the appropriate mountains being rounded or jagged.

In describing the scenario, of course we do have to employ concepts. If we are to fix on the scenario uniquely, we will indeed have to use very fine-grained concepts too, to capture the fine-grained content. But it is crucial to observe that the fact that a concept is used in fixing the scenario does *not* entail that that concept itself is somehow a component of the experience's representational content, nor that the concept must be possessed by the experiencer. The fine-grained concepts have done their work when they have fixed a unique spatial type. We should not confuse the scenario, the spatial type itself, with the infinitely various ways of picking it out.

Correlatively, we have on this account to recognize the rather indirect way in which descriptions in ordinary language, which are always at least partially conceptualized, help to characterize the way someone is experiencing the world. The ordinary-language characterization of the scenario can be at most partial.

(b) In some of my own earlier writings, I discussed the senses in which the type of content possessed by perceptual experience is analogue and unit-free.[5] Let us take "analogue" first. To say that the type of content in question is analogue is to make the following point. There are many dimensions—hue, shape, size, direction—such that any value on that dimension may enter the fine-grained content of an experience. In particular, an experience is not restricted in its range of possible contents to those points or ranges picked out by concepts—*red, square, straight ahead*—possessed by the perceiver. This fact is accommodated by attributing a scenario content. It is accommodated for characteristics of points in the environment because any values of a perceptible dimension may be mentioned in the ways which comprise the scenario. The restrictions on the environment determined by the veridicality of the experience need not be formulable using concepts possessed by the subject independently of the occurrence of the experience. With some important qualifications below, non-punctual properties such as shape in the scenario will be determined by the assignments to points; and again there will be no restrictions resulting from the thinker's repertoire of shape concepts on the shapes he may perceive things as having.

The unit-free nature of spatial perception is illustrated by the fact that when we see a table to have a certain width, we do not see it as having a certain width in inches, say, as opposed to centimeters. This is also explained by the distinction between the ways of characterizing a scenario and the scenario itself. Suppose we prescind here from qualifications about perceptual acuity. Then we can say that one and the same restriction on the distance between the sides of the table, one and the same restriction on the ways in which the space around the perceiver can be filled consistently with the experience being fully veridical, is given by doing these two things: saying that the sides are 39.4 inches apart, and saying that they are 100 centimeters apart.

The same point also holds for directions and the units in which they may be measured.

(c) The scenario account provides for the possibility of amodal contents of experience, in the sense that it allows for overlapping contents of experiences in different sense modalities. The restrictions on the environment required by the correctness of a visual experience may overlap with the restrictions required by the correctness of a tactile experience. This can be so when the positions, relative to each other, of the origins and axes for the scenarios of the contents of each experience are given in the subject's total conscious state. Both a visual experience, and a tactile experience resulting from stretching out your arm at a certain angle, may represent the existence of a surface in front of you at a certain distance. It is because this is so that, at a higher, concept-involving level, the judgement "That surface is red and warm" does not, when one is taking one's experience at face value, rest on any identity belief concerning the surface in question. I will consider amodal conceptual contents further below.

There is a very great deal more to be said on all aspects of this account. But even on this very primitive foundation, two considerations support the claim that an account mentioning scenarios cannot be dispensed with in favour of purely propositional accounts of the representational contents in question. By a "purely propositional" account, I mean one which identifies the representational content with a set of propositions (whether built to Frege's, Russell's, or some other specification), where the constituents of these propositions do not involve, directly or indirectly, scenarios.

First, it is hard to see how a purely propositional account can be made plausible without being parasitic on something properly treated by use of scenarios. Suppose you are in a field in the early autumn in England, and see a mist in a certain region. Can a theorist specify part of the representational content of your visual experience by means of the proposition, concerning that region, that it has the property of being misty? What is in question here is the Russellian proposition, which has the region itself as a constituent. This seems inadequate for very familiar reasons. Suppose the region in question is to your north. Someone for whom the region is in a north-easterly direction may also see it, that very same region, to be misty, and the same Russellian proposition would be used in specifying the content of his experience. But the region is clearly presented in perception in different ways to you and to the other person. Each of you sees it as being in a different direction relative to yourself, and your actions may differ as a result. Any description of the contents of your two experiences which omits this difference is incomplete. If we fill out the propositional theory to include "ways in which regions are perceived," the advocate of the scenario account will understandably say that these "ways" are a prime example of something of which only his account gives an adequate explanation.

To this the purely propositional theorist may reply that the relevant aspect of representational content should be formulated not just as "R is misty," where R is the region in question, but with a conjunction of Russellian propositions: "R is misty and R is located in direction D." Here D is an egocentric direction.

The theorist of scenarios should say that this move is inadequate on either of the two ways of taking it. The two ways of taking it result from two different ways of construing "egocentric." Take first the construal on which seeing something to be in egocentric direction D involves merely seeing it as having a certain direction in relation to object x, where in fact x is the perceiver himself. This reading is too weak to capture what is wanted. This is because one can see something as having a particular direction in relation to an object x which is in fact oneself whilst not realizing that the object to which one sees it as bearing that relation is in fact oneself. Examples of persons seen in mirrors suffice to make the point.

This suggests that the propositional theorist needs rather the stronger construal. On the stronger construal, to see something as having an egocentric spatial property is to see it as standing in a certain relation to oneself, where this involves use of the first-person way of thinking in giving the content of the visual experience. But the second consideration I wish to develop is precisely that purely propositional accounts, unlike the theory of scenarios, makes impossible an adequate account of the first-person way of thinking.

We have just seen that the pure propositionalist will have to mention the first-person way of thinking in giving his propositional contents of experience. For the pure propositionalist, propositional contents exhaust the nonconceptual representational content of experience. But this position is incompatible with the conjunction of two other principles which we have reason to accept.

One of these principles is what we can call "Evans's Thesis." This states that it is partially constitutive of a subject's employing the first-person way of thinking that he is prepared to make, noninferentially, suitable first-person spatial judgements on the basis of his perceptions when these are taken at face value. These will include "I am on a bridge" when he has an experience as of being on a bridge, "I am in front of a building," "There is a dog on my right," and so forth.

The other principle is what I call "the Principle of Dependence," which states that there can be no more to a concept than is determined by a correct account of what it is to possess that concept. If the Principle of Dependence is true, then we can (with many qualifications and refinements) individuate concepts by filling out this schematic form: concept F is that concept C to possess which a thinker must meet condition A(C). Here the concept F must not be mentioned as such within the scope of psychological states ascribed to the thinker. If it is, we will not have individuated the concept without residue in terms of what it is to possess it.[6]

In my judgement the Principle of Dependence is a powerful tool in the theory of thought and in philosophy more generally. But the application of the Principle which is of interest here is to first-person thought. If we accept Evans's Thesis, then the first-person way of thinking will be individuated in part by the rational sensitivity of present-tense spatial judgements containing it to the content of a thinker's perceptual experiences. But according to the propositionalist, these experiences themselves already have a first-person content which is not itself explained in terms of anything else. So the propositionalist will not have given an account of the first-person which respects the Principle of Dependence.

This problem is not solved for the pure propositionalist merely by saying that he holds that neither experience nor thought is prior in the individuation of the first-person way of thinking (a "no-priority"view). As long as it is agreed that part of the account of mastery of the first-person concept involves a certain distinctive sensitivity of first-person spatial judgements to the deliverances of perceptual experience, there is an obligation to say what that sensitivity is. If we accept the Principle of Dependence, an account of grasp of the first-person concept must distinguish it from all other concepts. Certainly the pure propositionalist does not have something individuating if he says that the first-person (for a given subject) is that concept C such that his judgements about whether C is F display a certain sensitivity to experiences which represent C as being F. This condition will be met by much else, including demonstrative ways of thinking of places in his immediate surroundings. It is of course open to a no-priority theorist to place further conditions on the concept C, relating to other aspects of first-person thought. But it is implausible that these have to be brought in to specify the kind of experience to which the sensitivity of first-person judgements is required. Intuitively, once we are given that a thinker is taking his experiences at face value, and asked whether he should rationally judge "There's a tree in front of me," we need to know only this: where the experience represents a tree as located in relation to the labelled axes of the relevant scenario.

The theory of scenarios, by contrast, provides a natural resource for solving this problem. It should not, of course, be any part of the scenario account to deny that the first-person has to be mentioned in fully specifying the representational content of perceptual experience. It should insist on first-person contents, as it should equally insist on the partially conceptual character of the perceptual content when one sees something to be a dog, or a tree. The issue is rather, in the case of the first person, whether or not the theorist can say more about what it is for the experience to have the first-person content. The scenario account can respect the Principle of Dependence applied to the first-person in the following way. The scenario account already says that a fundamental type of representational content is given by a scenario, a spatial type which involves a labelled origin and labelled axes. As the remarks of the preceding paragraph suggest, the rational sensitivity picked out in Evans's Thesis should be understood as a rational sensitivity of first-person present-tense-spatial judgements to the spatial relations things are represented in the scenario of the experience as having *to the labelled origin and axes*. This avoids the circularity, and in an intuitive way. Of course we still owe a philosophical account of what it is for one scenario, with one set of labelled axes and origin rather than another, to be the content of an experience. But once we recognize the level of the scenario, there is nothing to make this problem insoluble.

Devotees of the theory of indexical thought will note that points exactly corresponding to those just made about first-person thought can be made about the perceptual-indexical concept *here*. There is equally a circularity in the philosophical account of mastery of *here* in its relation to perceptual experience if we were not able to make reference to the labelled origin in an experience's scenario.

Despite these problems with the purely propositional approach, there is one range of phenomena of which the propositionalist may seem to have a superior treatment. These phenomena are experiences of "impossible" objects. The experience may be of an "impossible" depiction, or of an "impossible" three-dimensional object.[7] The problem such examples pose for the present account results from the fact that there is no way of filling in the space around the perceiver in the region of the "impossible" object which is consistent with the correctness of the experience: for, apparently, the experience has an inconsistent content. This problem is not satisfactorily resolved by leaving it unspecified, in such a case, what is in the region of the impossible object. That would not distinguish the experience of an "impossible" object from the quite different case of a subject who has a blind area, or scotoma, for that region.

Actually, the inconsistency of the properties of an "impossible" object emerge only with shifts of focal attention. The present treatment in terms of scenarios should be applied only to the focally attended region of space. If this is correct, it would not be surprising that shifts of attention are required for the impossibility of the object to emerge. The issue which then immediately arises is the content of the experience of the unattended regions. Cannot they still be inconsistent? However, what is distinctive of the unattended regions is that the various features perceived are not so precisely located. It is for this reason that focal attention is required if one is to see whether two different features perceived in an unattended region are possessed by the same object.

III. A FURTHER LEVEL OF CONTENT: AN APPLICATION

In the preceding section, we touched on one way in which scenarios can contribute to the individuation of a concept, the first-person concept. I turn now to discuss how some other conceptual contents are individuated in part by reference to a level of nonconceptual content. I will be suggesting that we need to recognize a kind of nonconceptual content which is distinct from that of the positioned scenario.

How is mastery of such apparently partially perceptual shape concepts as *square, cubic, diamond,* or *cylindrical* related to the level of nonconceptual content? In the general theoretical framework I favour, the task is to say how the various possession conditions for these concepts involve the level of nonconceptual content.[8] We can take the concept *square* as the example for discussion. The concept intended is the relatively observational shape concept which can be possessed without the subject's awareness of any geometrical definition. It is also a concept which has, inherently, fuzzy boundaries.

We can enter the issues by considering a natural, simple suggestion about what is necessary for possession of the concept *square*. The simple suggestion is built up using the materials we have developed so far. Suppose a thinker is taking his experiences at face value. Suppose too that in all the ways comprising the positioned scenario of his experience, the area of space apparently occupied by a perceived object is square. Then, this simple account suggests, the thinker must find the present-tense demonstrative thought that that object is square to be primitively compelling.

This simple account is not circular. It uses the concept *square* in fixing a certain sort of scenario; we emphasized earlier that that does not require the thinker to possess the concept *square*. This simple account can be written out in a way which makes it clearly capable of featuring as part of a longer story which, in the terminology of "What Are Concepts?" is of the A(C) form. Indeed it's plausible that any theory of possession of these relatively observational shape concepts will have at some point to exploit this way of avoiding circularity.

The necessary condition proposed by the simple account is, however, not in fact necessary. That it is not necessary is already shown by Mach's example of the square and the diamond:[9]

A thinker, taking his experiences at face value and possessing the concept *square*, need not find it primitively compelling (without further reflection) that a floor tile in the diamond orientation is square. But it can still be that in all the ways comprising the scenario of his experience, the region of space apparently occupied by the floor tile is square: as indeed it will be if his experience is veridical. The case illustrates a first respect in which we need to revise Evans' pioneering discussion of these issues. He wrote:

> To have the visual experience of four points of light arranged in a square amounts to no more than being in a complex informational state which embodies information about the egocentric location of those lights.[10]

For again, four points of light arranged in a regular diamond shape will produce an informational state which embodies information about the egocentric location of those lights. The informational state produced need not be an experience of them as arranged in a square.

Mach's example does not show that scenarios are irrelevant to the difference between the concepts *square* and *regular diamond.* What it does show is that they cannot be used in so simple a fashion; they need to be supplemented with the use of further materials. The correct way to do this, though, is not entirely straightforward.

Intuitively, the difference between perceiving something as a square and perceiving it as a (regular) diamond is a difference in which symmetries are perceived. When something is perceived as a diamond, the perceived symmetry is a symmetry about the bisectors of its angles. When something is perceived as a square, the perceived symmetry is a symmetry about the bisectors of its sides.[11] So intuitively, the simple account should be supplemented by requiring that, for the case it treats, the object apparently occupying the region in question be perceived as symmetrical about the bisectors of its sides. But does perceiving something as symmetrical require the perceiver to possess and make use of the concept *symmetrical?* If so, then what we will have done is simply to explain (this part of) mastery of the concept *square* in terms which require mastery of other concepts. We will not have anchored the difference between the concepts *square* and *diamond* at a level of nonconceptual content. If on the other hand there is a sense in which perceiving something as symmetrical does not require possession of the concept *symmetrical*, what is that sense? And how do we capture the nature of the additional content which distinguishes a case in which we have not merely veridical perception of a symmetrical region, but also have that symmetry perceived? I will take up these last two questions, in a speculative and programmatic way.

I suggest that perceptual experience has an additional layer of representational content. The contents in question cannot be identified with positioned scenarios, but they are nevertheless distinct too from conceptual contents. The contents of this additional layer I call *protopropositions*. These protopropositions are assessable as true or false. They contain properties or relations, rather than concepts of properties and relations. This is why I call them protopropositions rather than protothoughts. They contain such properties and relations as SQUARE, CURVED, PARALLEL TO, EQUIDISTANT FROM, SAME SHAPE AS, and SYMMETRICAL ABOUT.[12] These properties and relations can hold of places, lines or regions of the positioned scenario; or of objects perceived as located in such places. A protopropositions which is in the content of a perceptual experience will contain the objects, spatial or material, of which the property or relation is represented as holding, together with the property or relation represented to hold of it or them.

There will of course be many equally acceptable set-theoretic ways of building up such propositions. It is a necessary condition of an experience having the protopropositional content that an object is *F* that the object be, or occupy, a region of the experience's positioned scenario that really is *F*; and similarly for the relational contents of protopropositional contents. Though additional to scenario content, protopropositional content is inextricably rooted in scenario content.

Because it is properties and relations, rather than concepts thereof, which protopropositions contain, there is no immediate circularity in appealing to this level of representational content in individuating certain conceptual contents. As always, of course, we will want to have a substantive theory of what it is to be in perceptual states with protopropositional contents.

Protopropositional content is not determined by positioned scenario content. Two experiences can have the same positioned scenario content but different protopropositional contents. Many familiar cases can illustrate this. One illustration is given by the case in which one comes to see two shapes in the environment to be the same—one a rotation of the other, say, even though the single shape in question is quite unfamiliar. The experiences before and after seeing the identity of shape have the same representational content at the level of the positioned scenario. They differ in that the later experience has amongst its protopropositional contents that the two objects are of the same shape. An even more familiar case is given by certain differences in spatial grouping. When you see a two-dimensional array of elements as grouped in columns rather than rows, your experience has the protopropositional content that certain elements, which are in fact vertically arranged, are collinear. But there need not be any difference in the positioned scenario content between the case in which the array is seen as grouped in columns and that in which it is seen as grouped in rows.[13]

The reader will not be surprised to learn that it is this level of protopropositional content which I propose to employ to avert the threatened failure to give an account of the difference between the concepts *square* and *diamond*. When we say that in an experience of something as a square, the symmetry about the bisectors of its sides must be perceived, we are noting a restriction at the level of protopropositional content. An experience in which something is perceived as a square is one whose nonconceptual representational content contains the protoproposition that a certain figure is symmetrical about a line, a line which in fact in the positioned scenario of the experience bisects the figure's sides.

The reader may also have been wishing to protest for some time that there is an argument that the components of content I have tried to capture by the properties and relations of protopropositions are after all really conceptual. Consider the predicative component of the demonstrative content, judged on the basis of perception, "That line is straight." Isn't what is judged here of the line the same as the way experience represents the line to be? So if the component of the content judged is conceptual, must not the of the perception be likewise? I reply that there is indeed a conceptual component of the judgement that is intimately related to experiences with protopropositional contents that certain perceived things are straight. But this is the conceptual content which is individuated by means of certain of its relations to such protopropositional contents. As in the first-person case, and certain others to follow, it need not be part of the present position that perceptual experiences do not have conceptual contents involving this concept *straight*. What is part of the position is that if we are to have a noncircular account of mastery of this perceptual concept *straight,* that mastery must be related to some feature of experience which does not have to be explained in terms which presuppose possession of the concept. We cannot supply this by relating the mastery to experiences whose positioned scenarios require for their correctness that a certain line be straight: for that can be so without the straightness being perceived. Having the property STRAIGHT in the protopropositional content respects that point, without threatening a circular account of mastery.

The level of protopropositional content seems to me to merit further philosophical investigation. Besides the aspects already touched upon, two others can be mentioned. First, consider the difference between an experience in whose positioned scenario a region has a certain property, and one which additionally has in its protopropositional content that the region has that property. In the latter case, we seem to have the property that intuitively lies behind one case of the phenomenon Dretske calls "digitalization."[14] Further investigation of the case should lead to a better understanding of the phenomenon. Second, consider a subject who moves in the world. Suppose that the array of objects he perceives is itself stationary. The positioned scenario content of his experience will alter. But a protopropositional content to the effect that a line is straight, or two lines parallel, and so forth, will, if correct before the move, will be so after it too. So if he can keep track of the objects, lines or regions in question, we have here a very primitive—if not the most primitive—kind of perspective-independent content. A further investigation should say what grounds this perspective-independence; and say too how other perspective-independent notions are related to it.

Is the partial account of the perceptual concepts *square* and *diamond* which I have so far built up consistent with the amodality of these shape concepts? The natural criterion for the amodality of a concept within the present framework is this: no particular sense modality is such that experiences in that modality are required for a thinker to meet the concept's possession condition. Let us adopt that criterion. It treats amodality as a property of the concept itself, as opposed to a property dependent upon one amongst other means of grasping it. By this criterion, the modified simple account of the concept *square* can itself be a fragment of a possession condition for an amodal shape concept. Suppose a possession condition for a concept mentions a sensitivity of certain judgements to states with scenario-involving contents, or to anything else which can equally be present in more than one sense modality; and suppose too that it does not mention anything else which requires experiences in a particular modality. If these suppositions are met, then the concept is amodal. Despite the departure already made from it, this is very much in the spirit of Evans' idea. On the present account, there remains a "single conceptual capacity" which is exercised both in response to visual experience, and in response to tactile experience, when, in response to experiences of each kind, a thinker judges something to be square.[15]

Given the present criterion for the amodality of a concept, it is a *nonsequitur* to deny the amodality of a particular concept *F* solely on the ground that a subject can experience things as *F* in one modality but not in another. For it may be that in one modality, the protopropositional content of experience has, for that subject, properties that are, contingently, lacked by that level of content for the other modality. So, for example, symmetry and parallelism are not, as we actually are, made salient in the protopropositional content of our tactile experience in as wide a range of cases as they are in visual experience. Since this is so, some shape concepts may be applied on the basis of visual experience of an object which are not immediately applied when the object is touched. It does not follow that the concepts are modality-specific in the way the concept *red* is. We may rather be able to explain the difference by citing the different properties entering the protopropositional contents of visual and tactile experiences.

This treatment of perceptual shape concepts is consistent with acknowledging that experiences have conceptual content. In Evans' work, experiences are conceived as not having a conceptual content at all.[16] This part of Evans' conception is not obviously obligatory. It is not clear that there is good reason for denying the overwhelmingly plausible view that we see things as trees or hear a sound as that of a car approaching. However, accepting the overwhelmingly plausible view does not seem to give us a new resource to use in building possession conditions for perceptual shape concepts. If we try to make the possession condition for the perceptual concept *square* a matter of a certain sensitivity to experiences with a content containing the concept *square*, we will be open to the charge of circularity. This circularity is parallel to that suffered by the propositional theorist we noted earlier, who tries to make possession of the first-person concept a matter of the sensitivity of thoughts containing it to features of experiences which already have first-person contents. Like that theorist, we would be attempting to individuate a concept by reference to something which already takes for granted the thinker's possession of the concept in question, *viz.*, the capacity to have experiences whose contents contain the concept in question.[17]

The moral we should draw is that though experiences do indeed have conceptual contents, a possession condition for a concept should appeal at most only to the nonconceptual contents of the thinker's experiences, amongst which are the positioned scenarios and the protopropositional contents. If this is correct, then Evans' view that experiences have no conceptual contents was not by itself depriving him of any resource that is legitimately usable in the individuation of particular concepts.[18]

While the direction of constitutive explanation has so far been running from experience to concept possession, it is essential to allow too that there is some causal explanation in the opposite direction. Once a thinker has acquired a perceptually-individuated concept, his possession of that concept can causally influence what contents his experiences possess. If this were not so, we would be unable to account for differences which manifestly exist. One such difference, for example, is that between the experience of a perceiver completely unfamiliar with Cyrillic script seeing a sentence in that script, and the experience of one who understands a language written in that script. These two perceivers see the same shapes at the same positions; it can be that the positioned scenario and the protopropositional contents of their respective experiences are identical. The experiences differ in that the second perceiver recognizes the symbols as of particular kinds, and sequences of the symbols as of particular semantic kinds. The question of the nature of this difference, and more generally of what it is, constitutively, for an experience to have a conceptual content, remain as urgent and as open as ever.

Obviously the remarks of this section cannot purport to give a full possession condition for the concept *square*. A full possession condition must elucidate a thinker's ability to judge that an unperceived thing is square. Many intriguing issues arise in the attempt to elucidate that ability. One of them is the nature of a thinker's understanding that imperceptibly small things can be square, and that objects too large to perceive can be so too. Another such question is the relation of this understanding to the ability to judge of an unperceived but perceivable thing that it is square. Pursuing these issues would take us too far away from the other topics I wish to discuss. We can note, though, that the partial possession condition developed so far has a bearing on the answers to these questions. Consider, for instance, a thinker's ability to judge of something unperceived but perceptible that it is square. A theorist might try to explain this as follows. Just as the nonconceptual content of perception is given by a way of assigning objects, properties, relations and so forth in the space itself in which the subject is embedded. So far, this is tempting and plausible. But suppose the theorist goes on to say that to judge that an object at a specified, unperceived location is square is simply for the location of the content-specification of one's map to be assigned something square. That is identical with assigning it something diamond-shaped. But judging that the object there is square is different from judging it to be diamond-shaped. The distinctions between nonconceptual contents discussed above thus infuse our thought about the nonperceptual cases too.

IV. SPATIAL REASONING AND ACTION

If an experience enjoyed by a subject is to have a positioned scenario as its content, then the subject must have the capacity to draw on the content in a certain kind of spatial reasoning. The kind of spatial reasoning in question is that which manifests the subject's appreciation of the fact that a scene falling under the scenario presented in the experience is something to which his own spatial relations may vary. This will include as a special case that in which the scene is unperceived by him, when his spatial relation to it precludes his perceiving it. There can be conscious states whose properties are systematically explained by properties of a subject's surroundings, and indeed sequences of such states may explain the subject's behaviour. But if such states do not have contents which can be drawn upon in the kind of spatial reasoning just mentioned, then they do not have spatial significance for the subject himself.

The kind of spatial reasoning in question involves the subject building up a consistent conception of the world around him and of his location in it. I have been emphasizing that positioned scenario content is not conceptual content. So neither a model-theoretic nor a proof-theoretic notion of consistency employed for conceptual or propositional contents is precisely what we need in elaborating the notion of "a consistent conception" here. One of the distinctive relations spatial reasoning must exploit is rather the following notion: that of a perceptual experience's positioned scenario being consistent with a cognitive map. What this means is that there are ways of filling out the mapped space such that both of the following are instantiated: (i) the way of locating features in it required by the cognitive map and (ii) one of the ways of locating things in the world sufficient for the correctness of the positioned scenario.

Whenever we claim that a certain role in reasoning is essential to a particular kind of content, we incur an obligation. The obligation is to elucidate the relation between the content's role in reasoning and its correctness (truth) conditions (or contributions thereto). The obligation is one which, for logical constants, I made a stab at meeting in *Thoughts* and "Understanding Logical Constants."[19] It arises equally for spatial reasoning and scenario-involving contents. Ideally, we would like to have a theory of how a role in spatial reasoning determines the correctness of assigning one positioned scenario content rather than another to an experience. The task of providing this theory is one of the many which lie in intriguing regions visible from the main route of this paper. The task is of course not precisely analogous to that which arises for conceptual contents. For in those cases, the concepts are being individuated by their possession conditions; whereas, I have been emphasizing, a scenario (a spatial type) is not. Nonetheless, the obligation to say what it is for one spatial type rather than another to be involved in the content of an experience still exists, and must be met in part by detailed connections with spatial reasoning.

Genuine spatial reasoning must in turn be capable of explaining the spatial properties of a thinker's actions. The spatial property may be just the minimal property of being carried out at a particular location. The fact that such properties are included leaves room for the possibility of an equally minimal case, touched upon earlier. This is the case of a being which is not capable of initiating changes in its own configuration or location, but whose nonspatial actions (such as changes of colour or of the acidity of its surfaces) are controlled by its conception of its own and other things' locations on its cognitive map. Human beings, are, though, obviously capable of controlling a much wider range of the spatial properties of their actions, and it is worth considering the role of scenario content in this control. This further role cannot be taken as unconditionally constitutive of having experiences with scenario content, since the minimal case is possible. But the further role is crucial to the content-involving explanation of human action.

To characterize the further role, we must first remind ourselves of the distinctive kind of knowledge a subject has about the position of his limbs and the configuration of his body. As Wittgenstein and Anscombe long ago emphasized, this knowledge is not inferred from, or even caused by, sensations.[20] You can know the position of your own arm even when it is anaesthetized, and even when you are not seeing it or feeling it with another limb. What is important for us is that in the content of this distinctive kind of knowledge, the location of a limb is given egocentrically, in relation to the subject's body. It is given in the same kind of way in which a location is given in the positioned scenario of an experience, where the scenario is labelled with bodily axes. Further, it seems that, in the content of the distinctive knowledge, the limb whose position is known is not conceptualized—as a particular finger might be conceptualized as "the fourth finger on my left hand." On the way a particular finger is given in the content of the knowledge, it is potentially informative that that finger is the fourth on one's left hand. The knowledge, of a finger that is in fact the fourth on one's left hand, that it is in a certain location is something which can also be present in someone who has never thought of the finger as his fourth. The way the finger is given in the content of the distinctive knowledge seems to be one that features in the content of (sub)intentional states. Indeed it appears that the intention, or subintentional state, to move the finger thus given from one place to another is one which, when things are working properly, results in just that movement of the finer.

In this characterization of the distinctive kind of knowledge, I differ in some respects from Brian O'Shaughnessy's penetrating treatment.[21] O'Shaughnessy's insight was that knowledge of the position of our "bodily extremities" is "non-conceptual" and "non-propositional" (64). But he also says that this knowledge is "entirely practical"; and that its content "is exhaustively manifest in a set of physical acts" (64). It is certainly not true that a person's knowledge of the position of his limbs and body at a given time is exhaustively manifest in actions he actually performs. Much of this knowledge possessed at a given time is not put to use in action at all. Perhaps O'Shaughnessy would say rather that the distinctive knowledge consists in certain dispositions to act. There must be something right about mentioning dispositions to act, and I will return to them soon. But we will not get a proper understanding of the knowledge in question until we acknowledge its content. For it seems clear that the content of the distinctive knowledge has correctness conditions, which may or may not obtain. As a result of drugs or neural damage, a subject's belief about the location of his hand may be false. We cannot accommodate this just by saying that a false belief is one which, when manifested in action, does not lead to success. A distinction between success and failure requires a notion of the content of an intentional or subintentional state, a content whose correctness condition may obtain (success) or not (failure). And the content of these (sub)intentional states is of the same kind as the content of one's knowledge when one knows that one of them has succeeded.

It is not surprising, given his views, that O'Shaughnessy says that there is a structural parallel between subintentional states and sensations. If what I have been saying is right, the content of the subintentional state has a correctness condition on the external, objective world. The sensation does not. So there is no structural parallel of that sort on my position. Absence of conceptual content does not mean absence of all genuine content.

Now let us return to the further role of scenario content in human action. In supplying a subject with information about the location of things relative to bodily axes, perception supplies that nonconceptual information in a form immediately usable if the subject wants to move his body, or some limb, towards, or from, or to be in some other spatial relation to what he perceives. It is only because the bodily-based scenario-involvingcontents are common to perception, to our distinctive knowledge of bodily position, and to what is under our immediate (sub)intentional control, that the information in perception is immediately usable in performing actions.

This intermediary role of scenario contents with labelled (bodily) axes explains more fully a connection, which I mentioned in earlier work, between the nonconceptual contents of perception and bodily action. In "Analogue Content" and "Perceptual content," I noted that a normal individual, asked to direct the beam of a spotlight in a forest onto a tree 47 degrees to the right of straight ahead, will not, in the sense relevant to action, know where to point the beam. The normal individual does not know which perceptually individuated direction is 47 degrees to the right of straight ahead. The normal individual will, though, have no difficulty if the tree in question is marked, so that he can see in which precise direction it lies.

He has no difficulty in this second case, because now the tree is represented as being in a certain direction in the positioned scenario of his current perception. The nonconceptual content of the individual's knowledge of the positions of his limbs, which have to be moved to move the spotlight, also gives their location relative to the same bodily axes as are used to specify the scenario presented in his perception of the tree. These two facts can then engage to produce appropriate action because the scenario-involving indication of direction given in the perceptual state can feature in basic (sub)intentional states—that is, in states which the subject does not plan to implement by doing something else. When an intentional or subintentional state is basic, and all goes properly, the instructions to carry it out lead to the computation and execution of the appropriate bodily movements without the further instructions or information from the personal level. It is this combination of facts which allows the individual, in the second case, to move the beam of the spotlight to point in the perceived direction of the tree, without further information.

There are unusual cases in which two or more systems of labelled axes are properly used in giving the positioned scenario content of a single experience, but in which the spatial relation between those two systems is not specified in the content of that experience. There are, for instance, certain positions in which, when lying on your back, you can twist your arms up behind your head. A piece of furniture touched with a hand of the twisted arm may be experienced as standing in certain spatial relations to one's hand, fingers and wrist, but not as standing in any particular relation to the rest of one's body. In such a case of fragmentation, there is also fragmentation of the person's systems of knowledge which help to explain bodily actions. The subject will know what to do to move his fingers away from the touched furniture, but not, without further reasoning, how to move them closer to his torso.

Of the many substantive questions which arise in this area, one which seems somewhat more tractable than the rest is this: what gives a subintentional state or trying one nonconceptual content rather than another? One subpart of this question concerns the labelling of axes in a scenario. We need to say something illuminating about this labelling which does not take for granted a partially scenario-like nonconceptual content which already has labelled axes.

I should emphasize that the problem here does not concern merely notational variants of particular axes and coordinate systems. One notational variant of the basic human axes would rotate them by 45 degrees, and adjust coordinates in specifying a scenario in a compensating fashion. That this is a purely notational, insubstantial, matter should be especially clear in the case of scenario content—for it is clear here that the variation is solely in means of specifying one and the same spatial type. The important question is rather this: why are axes labelled in *some* way involving particular bodily parts and limbs appropriate in giving the spatial type which is in turn used in giving the content of a trying, of other subintentional states? We could encapsulate the point by saying that the question is about frames of reference for spatial types, rather than about coordinate systems.

Here is one possible answer. Let me say that two instructions "tryings" of a given subject are of the same type if they differ only in the reference of their "now" component. So trying to move one's left hand to a particular position in front of oneself at 9 a.m. and trying to do so at 10 a.m. are instructions of the same type. Now take a given type of instruction with a nonconceptual, partially scenario-like content. I suggest that the frame of reference to be used in labelling the axes of its scenario is that frame with respect to which instructions of the given type always have the *same* effect, when characterized in relation to that frame, in all normal spatio-temporal contexts (and when the efferent nervous system is functioning properly). When an instruction is an instruction to move one's hand in a certain relation to one's body, it will have the same effect, described in relation to a bodily frame of reference, whether the thinker is in London or in Edinburgh. It will not have the same effect if the bizarre choice were made of a frame of reference involving longitude and latitude. Nor would it always have the same effect (when the thinker is functioning properly) in any other frame of reference defined by objects with respect to which the subject can move. Nor will descriptions of the effect in terms of muscle-changes be the same in all spatio-temporal contexts. For they will depend on the starting point of the hand.

This criterion can be called the *constancy* criterion for fixing the frame of reference, and so the labelling, of any axes used in individuating the scenarios for a given subject's subintentional states. We can note that the constancy criterion cuts below the level of content—even below the level of scenario content—and so is of the right form, at least, to contribute to a substantive theory of subintentional states with nonconceptual contents.

Before closing, I wish to touch very briefly upon a few of the links between the types of nonconceptual representational content I have been discussing and issues in the cognitive sciences. Suppose, with Roger Shepard, we regard the task of the mechanisms of visual perception as that of computing an inverse of the projective mapping from the environment to the retinal image.[22] From this lofty perspective, we would certainly expect representations with positioned scenario contents to be computed (along the way): for it is the real scene, in our sense, which produces the retinal image. In a content-involving psychology, we would expect the early stages of vision to compute and combine various partial specifications of the scenario instantiated around the perceiver. The several feature maps, and the integration thereof, studied by Anne Treisman are just such partial specifications of scenarios.[23] They also begin to suggest mechanisms by which protopropositional content is made explicit in mental representations. If we want the explanatory power that only a content-involving description of a computation can supply, we should use positioned scenarios and protopropositions in describing the mechanisms outlined in that work.

States with scenario contents also seem particularly well-suited to receive a realization in a connectionist network. We can partition the three-dimensional space around the subject into suitably small cells. Each cell could be represented by a "binding" unit which is connected to three other elements. These three other elements each represent the values of the cell on the labelled axes of the scenario. The binding unit can then have connections to other assemblies for the features represented as instantiated at that cell; and so forth. More generally, I suggest that positioned scenario content is one fundamental kind of content appropriate to the mental models discussed in philosophical terms by Colin McGinn.[24] I will refrain from further speculation, but draw this general moral: that here, as elsewhere, our future investigations and hypotheses should make essential use of the dual ideas of positioned scenarios and of protopropositional content.

NOTES

1. Earlier versions of this material were presented in 1988-9 to an interdisciplinary conference on the mental representation of space at King's College Cambridge, and to a seminar at Oxford University. I thank John Campbell, Adrian Cussins, Martin Davies and Michael Martin for valuable comments.

2. See T. Burge, "Demonstrative Constructions, Reference, and Truth," *Journal of Philosophy* LXXI (1974) 205-223.

3. There are several oversimplifications here; I am aiming to capture the spirit of a position.

4. For these two ways of representing surface orientation at a point, see D. Marr, *Vision* (San Francisco: Freeman, 1982), pp. 241-3.

5. "Analogue Content," *Proceedings of the Aristotelian Society* Supplementary Volume LX (1986) 1-17; and "Perceptual content," forthcoming in *Themes from Kaplan*, ed. J. Almog, J. Perry and H. Wettstein (New York: Oxford University Press).

6. See "What are Concepts?" in *Midwest Studies in Philosophy 1989: Contemporary Perspectives in the Philosophy of Language II,* ed. P. French, T. Uehling and H. Wettstein.

7. See the photograph in R. Gregory's *The Intelligent Eye* (London: Weidenfeld and Nicolson, 1970), p. 55.

8. See "What Are Concepts?" for further elaboration of the framework.

9. E. Mach, *The Analysis of Sensations* (Chicago: Open Court, 1914), P. 106.

10. *Ibid.* p. 392.

11. "People perceive different geometrical properties of the figure when it is seen as a square rather than as a diamond": S. Palmer "The Psychology of Perceptual Organization: A Transformational Approach," p. 293 in *Human and Machine Vision* (Academic Press: New York, 1983), ed. J. Beck and Rosenfeld volume. Palmer's paper contains further material of great relevance to a philosophical theory of these matters.

12. I use upper case letters for a word to indicate that I am referring to the property or relation to which it refers, rather than to the concept it expresses.

13. The preceding discussion revises the treatment of this case given in my *Sense and Content* (Oxford: Clarendon Press, 1983), without altering the general view required of the relation between sensational and representational properties of experience.

14. I speak of the underlying intuition, rather than Dretske's own definition of the notion. For Dretske's own discussion, see chapters 6 and 7 of his *knowledge and the Flow of Information* (Cambridge, Mass.: MIT Press, 1981).

15. *Collected Papers* p. 374.

16. See *The Varieties of Reference* (Oxford: Clarendon Press, 1982); the view is endorsed by C. McGinn *Mental Content* (Oxford: Blackwell, 1989), p. 62.

17. This was a problem which was noted, but far from satisfactorily resolved, in my *Sense and Content* (Oxford: Clarendon Press, 1983).

18. John Campbell has emphasized the philosophical significance of real cases of subjects who are able to identify and apparently perceive the shape of objects in their environment, without experiencing them as having any particular (egocentric) location. (See his review, in the *Journal of Philosophy* LXXXVI (1989), of G. Evans' *Collected Papers:* and the references to the psychological literature included in his review.) As Campbell says, these cases show another respect in which we need to revise Evans' description, displayed above, of perceiving something as square. For Evans' description apparently leaves no room for the possibility of experiencing something as square without experiencing it as having a particular location. The account we have built up so far at least formally leaves room for the possibility. Consider a case in which something is experienced as square, but is not experienced as possessing any particular location. The positioned scenario of the subject's experience is then as follows. In each of the ways comprising that scenario, there is something square filling a region of egocentric space; but it is not the same region for all of the different ways comprising the scenario. (The distinctive perceived symmetry at the level of protopropositional content which is required for experiencing something as square is consistent with this description.) Of course we need a substantive psychological and philosophical theory to explain how these extraordinary cases are possible.

19. *Thoughts: an Essay on Content* (Oxford: Blackwell, 1986), and "Understanding Logical constants: A Realist's Account" in *Proceedings of the British Academy* LXXIII (1987) 153-200.

20. L. Wittgenstein, *Philosophy of Psychology* (Oxford: Blackwell, 1980), for instance in Vol. I, sections 770-2, 798; G. E. M. Anscombe, "On Sensations of Position" repr. in her *Metaphysics and the Philosophy of Mind: Collected Papers Volume II* (Minneapolis: University of Minnesota Press, 1981).

21. See his chapter "The Subintentional Act" in volume II of *The Will* (Cambridge: CUP, 1980).

22. See his "Psychophysical Complementarity," in *Perceptual Organization*, ed. M. Kubovy and J. Pomerantz (Hillsdale, NJ: 1981).

23. A. Treisman and G. Gelade "A Feature-Integration Theory of Attention," *Cognitive Psychology* 12 (1980) 97-136; and A. Treisman and H. Schmidt "Illusory Conjunctions in the Perception of Objects," *Cognitive Psychology* 14 (1982) 107-141.

24. See his *Mental Content* (Oxford: Blackwell, 1989), Part 3. McGinn indexes mental models with truth-evaluable propositions. If these propositions are built up from constituents solely at the level of reference, I would remarshal the arguments of Section 2 of this paper to make a case that mental models deserve a more discriminating assignment of content. If the indexing propositions are at the level of sense, then I would say that their constituents cannot be explained philosophically without mentioning scenario content. The assignment of positioned scenarios as contents of mental models is entirely within the spirit of McGinn's views.

PHILOSOPHICAL STUDIES SERIES

Founded by Wilfrid S. Sellars and Keith Lehrer

1. JAY F. ROSENBERG, *Linguistic Representation,* 1974.
2. WILFRID SELLARS, *Essays in Philosophy and Its History,* 1974.
3. DICKINSON S. MILLER, *Philosophical Analysis and Human Welfare*. Selected Essays and Chapters from Six Decades. Edited with an Introduction by Lloyd D. Easton, 1975.
4. KEITH LEHRER (ed.), *Analysis and Metaphysics*. Essays in Honor of R. M. Chisholm. 1975.
5. CARL GINET, *Knowledge, Perception, and Memory,* 1975.
6. PETER H. HARE and EDWARD H. MADDEN, *Causing, Perceiving and Believing*. An Examination of the Philosophy of C. J. Ducasse, 1975.
7. HECTOR-NERI CASTAÑEDA, *Thinking and Doing*. The Philosophical Foundations of Institutions, 1975.
8. JOHN L. POLLOCK, *Subjunctive Reasoning,* 1976.
9. BRUCE AUNE, *Reason and Action,* 1977.
10. GEORGE SCHLESINGER, *Religion and Scientific Method,* 1977.
11. YIRMIAHU YOVEL (ed.), *Philosophy of History and Action*. Papers presented at the first Jerusalem Philosophical Encounter, December 1974, 1978.
12. JOSEPH C. PITT, *The Philosophy of Wilfrid Sellars: Queries and Extensions,* 1978.
13. ALVIN I. GOLDMAN and JAEGWON KIM, *Values and Morals*. Essays in Honor of William Frankena, Charles Stevenson, and Richard Brandt, 1978.
14. MICHAEL J. LOUX, *Substance and Attribute*. A Study in Ontology, 1978.
15. ERNEST SOSA (ed.), *The Philosophy of Nicholas Rescher: Discussion and Replies,* 1979.
16. JEFFRIE G. MURPHY, *Retribution, Justice, and Therapy*. Essays in the Philosophy of Law, 1979.
17. GEORGE S. PAPPAS, *Justification and Knowledge: New Studies in Epistemology,* 1979.
18. JAMES W. CORNMAN, *Skepticism, Justification, and Explanation,* 1980.
19. PETER VAN INWAGEN, *Time and Cause*. Essays presented to Richard Taylor, 1980.
20. DONALD NUTE, *Topics in Conditional Logic*, 1980.
21. RISTO HILPINEN (ed.), *Rationality in Science,* 1980.
22. GEORGES DICKER, *Perceptual Knowledge,* 1980.

23. JAY F. ROSENBERG, *One World and Our Knowledge of It,* 1980.
24. KEITH LEHRER and CARL WAGNER, *Rational Consensus in Science and Society,* 1981.
25. DAVID O'CONNOR, *The Metaphysics of G. E. Moore,* 1982.
26. JOHN D. HODSON, *The Ethics of Legal Coercion,* 1983.
27. ROBERT J. RICHMAN, *God, Free Will, and Morality,* 1983.
28. TERENCE PENELHUM, *God and Skepticism,* 1983.
29. JAMES BOGEN and JAMES E. McGUIRE (eds.), *How Things Are, Studies in Predication and the History of Philosophy of Science,* 1985.
30. CLEMENT DORE, *Theism,* 1984.
31. THOMAS L. CARSON, *The Status of Morality,* 1984.
32. MICHAEL J. WHITE, *Agency and Integrality,* 1985.
33. DONALD F. GUSTAFSON, *Intention and Agency,* 1986.
34. PAUL K. MOSER, *Empirical Justification.* 1985.
35. FRED FELDMAN, *Doing the Best We Can,* 1986.
36. G. W. FITCH, *Naming and Believing,* 1987.
37. TERRY PENNER, *The Ascent from Nominalism.* Some Existence Arguments in Plato's Middle Dialogues, 1987.
38. ROBERT G. MEYERS, *The Likelihood of Knowledge,* 1988.
39. DAVID F. AUSTIN, *Philosophical Analysis.* A Defense by Example, 1988.
40. STUART SILVERS, *Rerepresentation.* Essays in the Philosophy of Mental Rerepresentation, 1988.
41. MICHAEL P. LEVINE, *Hume and the Problem of Miracles. A Solution,* 1979.
42. MELVIN DALGARNO and ERIC MATTHEWS, *The Philosophy of Thomas Reid,* 1989.
43. KENNETH R. WESTPHAL, *Hegel's Epistemological Realism.* A Study of the Aim and Method of Hegel's *Phenomenology of Spirit,* 1989.
44. JOHN W. BENDER, *The Current State of the Coherence Theory.* Critical Essays on the Epistemic Theories of Keith Lehrer and Laurence Bonjour, with Replies, 1989.
45. ROGER D. GALLIE, *Thomas Reid and 'The Way of Ideas'*, 1989.
46. J-C. SMITH (ed.), *Historical Foundations of Cognitive Science,* 1990.
47. JOHN HEIL (ed.), *Cause, Mind, and Reality.* Essays Honoring C. B. Martin, 1990.
48. MICHAEL D. ROTH and GLENN ROSS (eds.), *Doubting.* Contemporary Perspectives on Skepticism, 1990.
49. ROD BERTOLET, *What is Said.* A Theory of Indirect Speech Reports. 1990
50. BRUCE RUSSELL (ed.), *Freedom, Rights and Pornography.* A Collection of Papers by Fred R. Berger. 1990
51. KEVIN MULLIGAN (ed.), *Language, Truth and Ontology.* 1992
52. JESÚS EZQUERRO and JESÚS M. LARRAZABAL (eds.), *Cognition, Semantics and Philosophy.* Proceedings of the First International Colloquium on Cognitive Science, 1992

VOLUME 52